Engineering Practical Chemistry

ENGINEERING PRACTICAL CHEMISTRY

[For the Students of B.E./B.Tech./B.Arch./ B.Sc.(Engg.)/B.Sc./Polytechnic Courses]

By

DR. MANAS RANJAN SENAPATI

M.Sc., M.Phil., PhD, FNASc.

Professor and Head Department of Chemistry

NM Institute of Engineering and Technology, Bhubaneswar

Odisha

UNIVERSITY SCIENCE PRESS

(An Imprint of Laxmi Publications Pvt. Ltd.)

BANGALORE • CHENNAI • COCHIN • GUWAHATI • HYDERABAD

JALANDHAR • KOLKATA • LUCKNOW • MUMBAI • RANCHI

NEW DELHI

Published by :
UNIVERSITY SCIENCE PRESS
(*An Imprint of Laxmi Publications Pvt. Ltd.*)
113, Golden House, Daryaganj,
New Delhi-110002
Phone : 011-43 53 25 00
Fax : 011-43 53 25 28
www.laxmipublications.com
info@laxmipublications.com

First Edition : 2011

OFFICES

✆ **Bangalore**	080-26 61 15 61	✆ **Jalandhar**	0181-222 12 72
✆ **Chennai**	044-24 34 47 26	✆ **Kolkata**	033-22 27 43 84
✆ **Cochin**	0484-237 70 04, 405 13 03	✆ **Lucknow**	0522-220 95 78
✆ **Guwahati**	0361-254 36 69, 251 38 81	✆ **Mumbai**	022-24 91 54 15, 24 92 78 69
✆ **Hyderabad**	040-24 65 23 33	✆ **Ranchi**	0651-221 47 64

UEP-9539-100-ENGG. PRACTICAL CHEM-SEN. **C—**

Typeset at : Goswami Associates, Delhi. *Printed at* : Ajit Printers

Dedicated to My Mother

CONTENTS

PREFACE

The textbook **"Engineering Practical Chemistry"** is intended for the students of B.E./B.Tech./B.Arch. of all Indian Universities (Anna, BPUT, JNTU, MDU, NMU, PTU, Shivaji, SVU, UPTU, WBUT, etc.). The book is also useful equally for B.Sc. and polytechnic students.

In order to make the book of more universal use standard methods such as those given in the "methods of analysis of American Society for Testing Materials" and the "methods of analysis of the association of official agricultural chemists" have in many cases been introduced.

Special methods have been included that have been found to be saving of both time and labour. The author has made an effort to provide conceptual understanding of the principles involved, before the student attempts to perform the experiment. A number of questions have been given at the end of each chapter, which will help the students for appearing their viva-voce in practical examination. It is observed that students face difficulty to solve the problems based on experiments. Some solved numerical problems are given at the end for practice of students. Finally a number of important appendices have been added to make the book informative.

In sincerely convey my appreciation and thanks to all my colleagues for their suggestion and timely help. I also record my sense of appreciation to my students whose questions and interest during my laboratory classes provided valuable input into this effort. My thanks are due to my family members who have been constantly inspiring me which worked as a catalyst behind my success.

Any suggestions for improving the book will be gratefully received by the author.

—AUTHOR

Chapter 1 GENERAL INSTRUCTIONS

1.1 LABORATORY RULES

Safety in the chemistry laboratory is very important to avoid most accidents in the laboratory which generally occur because of carelessness of worker. Presence of teachers and demonstrators in laboratory is essential. It is also essential for the students to know proper handling of the apparatus and chemicals. Students must be very careful and responsible while doing the experiment. Following are certain rules to prevent most of the accidents.

1. Use safety glasses during the laboratory period.
2. Always wear a laboratory apron during the laboratory period.
3. Never work alone in the laboratory.
4. Do not use any cracked or broken glasswares.
5. Note the place of fire extinguisher and know the method of using it.
6. Keep the reagent bottles at their proper places after use.
7. Read the label carefully on the reagent bottle before using.
8. Never use reagents from the unlabelled bottle.
9. Never taste any chemical as most of them are injurious and toxic in nature.
10. Keep a distance while observing odour of a substance/gas.
11. Do not throw waste such as match stems, filter papers etc., into the sink.
12. Avoid direct contact of solvents/chemicals with the skin.
13. Check the leakage of gas before lighting the burner.
14. Follow the directions of your demonstrator or teaching assistant before doing any experiment.

1.2 LABORATORY FIRST-AID

Each laboratory must have a well equipped **First Aid Box** ready for use. Even with best precautions, sometimes an accident may occur. In this case, act as follows:

(*a*) **Fire:** Bench fires may be extinguished either by fire extinguisher or by sand. To help prevent spread of the fire, remove any containers of flammable solvent from the immediate area and turn off any burners in the fire affected area. It is advised not to get nervous during such an incident.

(*i*) *Burning Clothing:* If your clothing is on fire, do not run because rapid movement will only fan the flames. Either remove the burning cloth if possible or lie down on the floor. Your neighbours can help to extinguish the flames by using fire blankets, laboratory aprons etc., that are immediately available.

(*ii*) *Burning Reagents/Chemicals:* Switch off the gas line and also hot plates. If the fire is in a beaker or flask, cover the mouth so that fire will extinguish due to absence of air. If fire has spread then use fire extinguisher.

(*b*) **Burns**

(*i*) *Alkali Burn:* Wash immediately with excess of water, then with 1% acetic acid and finally with water. Allow to dry, apply "silverex" (antimicrobial cream) and consult doctor.

(*ii*) *Acid Burn:* Wash with plenty of water, then with very dilute sodium bicarbonate solution and finally with water to remove the acid.

(*iii*) *Bromine Burn:* Soak bromine with tissue paper, wet thoroughly with water, apply glycerol to the affected part. After sometime remove glycerol by washing and apply "silverex".

(*iv*) *Sodium Burn:* Remove any piece of sodium with forceps and put in kerosene oil. Wash thoroughly the affected part with water, then with 1% acetic acid and finally with water. Allow to dry, apply "Silverex".

(*c*) **Cuts:** Minor cuts must be washed with plenty of water and be treated by physician. If bleeding is severe, apply direct pressure around the wound and consult doctor.

(*d*) **Chemical in Eyes:** Wash the eye with plenty of water to remove most of foreign matter. Keep eye open and wash with cold, clean water several times. Consult the doctor.

1.3 CLEANING OF GLASSWARE

The classic glassware cleaner is chromic acid. This is prepared as follows. In 1 litre of H_2O dissolve 100 g of $K_2Cr_2O_7$. Add slowly, with great care and while stirring 1 litre of concentrated H_2SO_4. Use hot or allow the glassware to soak overnight. This mixture is hygroscopic and caustic so containers should be covered. Glasswares may also be washed with ordinary detergents and thoroughly washed.

Washed apparatus is allowed to drain and finally may be rinsed with acetone dried in air or in hot air oven. Graduated glassware should not be dried in a hot air oven because its calibration may be spoiled.

1.4 REAGENTS AND CHEMICALS

Reagents and chemicals must be of high purity, chemicals of analytical reagent quality (AR) and distilled water (IS 1070–1960) are used in preparing the solutions of reagents. The AR grade of purity is not always necessary. The Cheaper Laboratory Grade may then be used.

Making up Solutions: Where two liquids are to be mixed it may be best to add the larger volume to the smaller one because most of the mixing is effected during the addition of the two liquids. But with H_2SO_4 the heat of dilution is so great that the solution may boil violently and cause drops of liquid to be thrown out. H_2SO_4 should be diluted by adding the acid slowly, carefully and with stirring to water.

Where a solid and liquid are to be mixed it is usually best to add the solid to the liquid, stirring the while because the solid then disperses. Readily soluble solids may otherwise form a 'cake' with the first drops of liqiud , and then be difficult to disperse. It may occasionally be useful to make use of the heat of solution to aid dissolution by adding the solid to only a small part of the liquid.

Mixing must always be thorough and careful.

1.5 HANDLING REAGENTS

Reagents should be placed in a systematic manner so that they may be readily available. It is important to keep the reagent bottles in their respective place after use. Reagent bottles should be kept stoppered except when they are in use. Reagent bottles must be labelled. Do not pour any reagent directly from a bottle into strongly acidic or basic solutions or into hot solutions.

1.6 MAINTENANCE OF RECORD

A practical record is to be maintained by the student giving theory, understanding over the experiment arriving at the conclusion. Evaluation for an experiment is based on the presentation of record.

A reporting format is being made as below which may have slight changes depending upon the experiments.

Left Hand Page

1. Chemical equations
2. Diagram/Flow sheet
3. Weights taken
4. Titration figures
5. Graph.

Right Hand Page

1. Experiment number and date.
2. Aim of the experiment (Write clearly the experiment alloted).
3. Apparatus required.
4. Chemicals required.
5. Theory (Give an account of the theory and formula to be used).
6. Procedure (Give the detail of how you have actually performed the experiment).
7. Tabulation observations (Note down the readings directly on the copy. No overwriting is permitted in the observation table).
8. Calculation (Perform all calculations clearly).
9. Conclusion (Give in the form of object).
10. Precautions (Mention all necessary precautions to be taken).

Chapter 2

LABORATORY EQUIPMENTS AND TECHNIQUES

2.1 GLASS APPARATUS

All the glass ware used in the laboratory should be made of heat resistant glass like pyrex which can withstand high temperatures, as and when necessary. Some of the commonly used glass ware are discussed in detail.

Flasks

The commonly used flasks are shown in Fig. 2.1. They find use in refluxing and distillation. Type (*vi*) is the common conical flask (also called as Erlenmeyer flask) available in size from 5 ml to 6 litres with or without graduation used for mixing and titration.

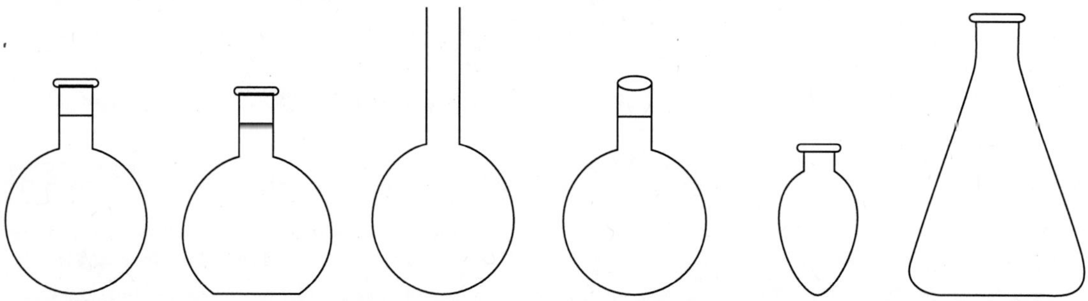

(i) (ii) (iii) (iv) (v) (vi)

FIGURE 2.1

Volumetric Flask

Volumetric flasks have a flat bottom, pear-shaped body and, a long narrow neck. An air tight glass stopper is fitted at its mouth which prevents leakage of liquid while shaking. At the neck of the flask a mark is engraved. When the liquid is filled upto this mark, the volume of the liquid equals to the volume printed over the flask. The flasks are generally available from 5 ml to 2 litres. They find use in the preparation of standard solutions. Some commonly used volumetric flasks are shown in Fig. 2.2.

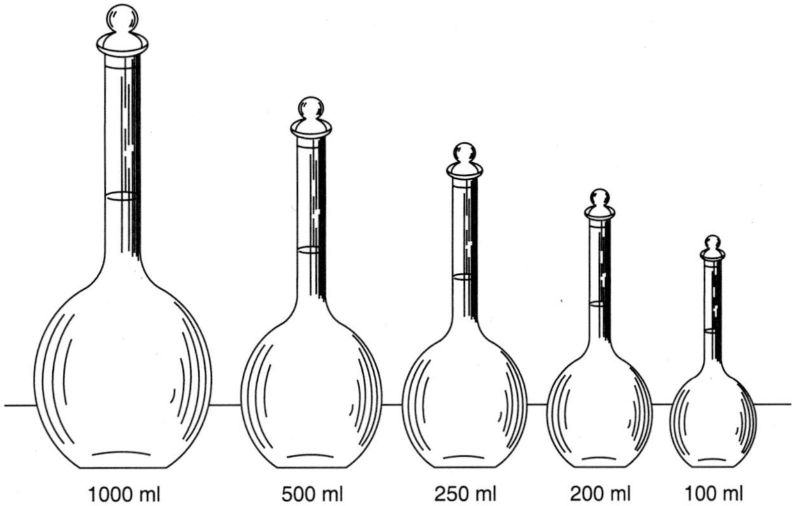

FIGURE 2.2 **Volumetric flask.**

Funnels

The commonly used funnels are shown in Fig. 2.3. Type (*i*) shows an ordinary funnel, whereas (*ii*) and (*iii*) are the separatory funnels used for extraction of a product from the reaction mixture. Type (*iv*) is a dropping funnel employed for the addition of a reactant to a reaction mixture.

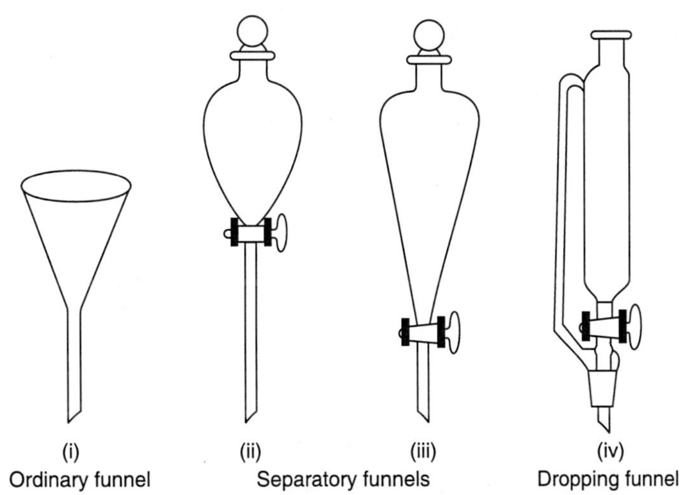

FIGURE 2.3 **Funnels.**

The Buchner funnel and the Buchner flask (Fig. 2.4) are used extensively in synthesis for filtration of a solid product, which is usually difficult to be filtered, by any other means.

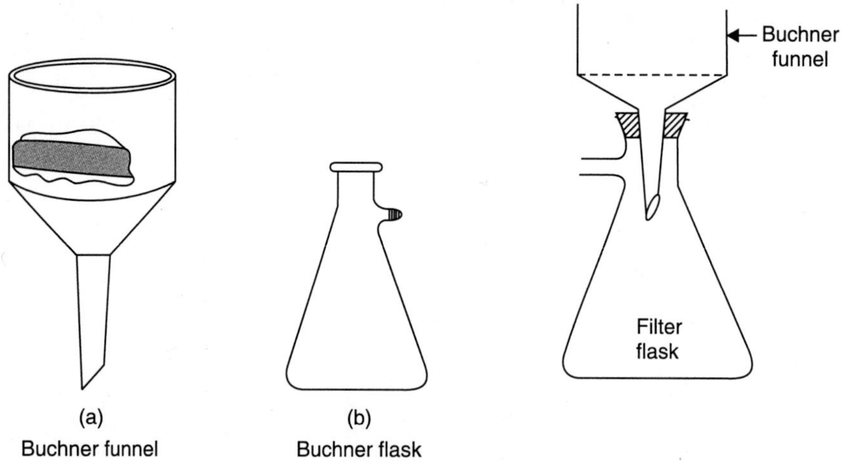

FIGURE 2.4

Pipette

Pipette is a slender graduated tube with a bulb in the middle for measuring and transferring liquids from one vessel to another. When a pipette is filled up with a liquid upto the mark on its stem, volume of the liquid transferred is equal to the volume printed on the bulb of the pipette. Graduated pipettes are also available for transferring out smaller volumes of solutions (Fig. 2.5).

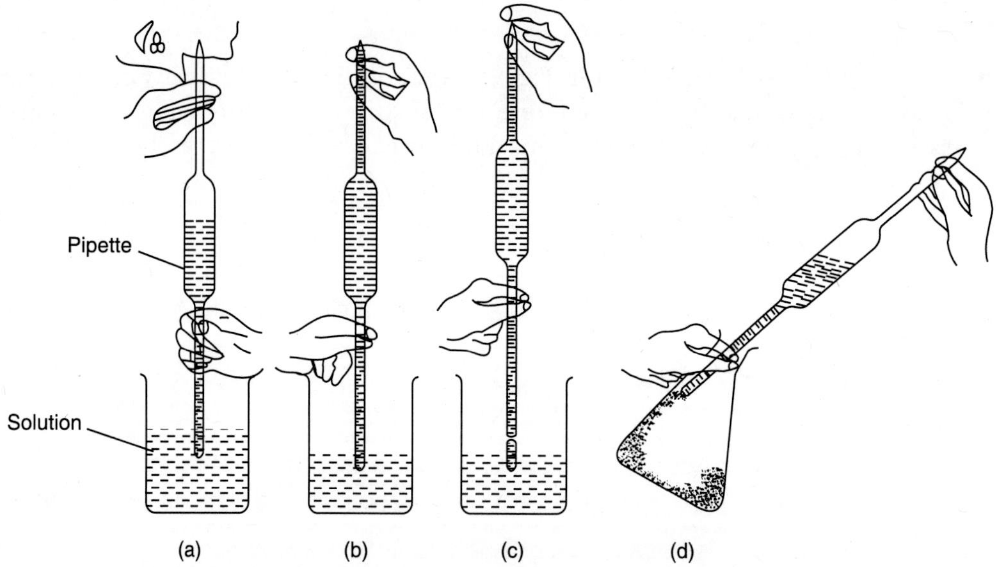

FIGURE 2.5 **(*a*), (*b*) (*c*) and (*d*), measuring of solution with a pipette.**

Before filling the pipette with standard solution wash thoroughly and then rinse twice with small amount of solution to be sucked. Dip the nozzle of the pipette into the solution and suck 1 ml. above the mark. Put immediately the fore-finger tip on the top of the pipette. Hold the pipette vertically and gently release the pressure without removing the finger until the lower meniscus of the solution comes in level with the mark on the tube of the pipette. Transfer the solution in the flask and remove the last drop sticking to the nozzle of the pipette by touching it to the wall of the flask. Do not blow to expel the last drop.

Burette

Burette is a graduated glass tube with a stop cock or pinch cock at the bottom (Fig. 2.6). The commonly used burette is of 50 ml. capacity.

Before use, the burette, must be perfectly clean. Rinse the burette with small volume of the standard solution twice. Fill it with the standard solution. Remove air bubble if present. Always read lower meniscus of the liquid keeping your eye at the same level of meniscus. While taking reading, keep the eye at the same level of the meniscus and as close to the eye as possible to avoid error due to parallax.

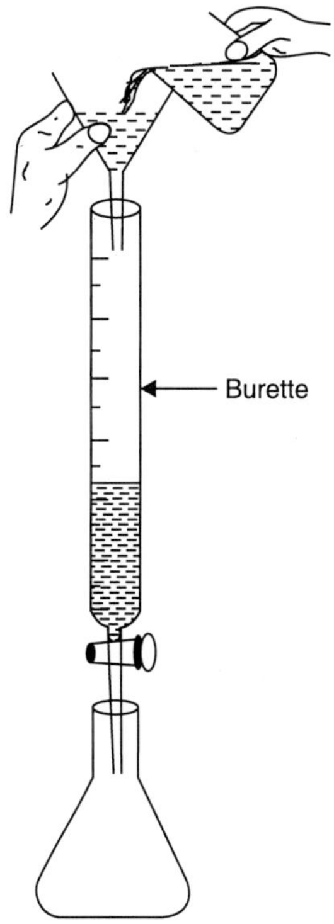

FIGURE 2.6 **Burette being used in titration.**

Measuring Cylinder

These are graduated vessels available in capacities from 1 ml to 2000 ml. Since the area of the surface of the liquid is much greater than in a graduated flask, the accuracy is not very perfect. They find use where only rough measurements are required.

2.2 HEATING APPARATUS

Bunsen Burner

An ordinary bunsen burner is widely employed for the attainment of moderately high temperatures. The supply of air may be regulated by air holes at the base and a clean non-luminous flame may be obtained.

Meker Burner

A meker burner has a large hole for admitting air freely so that the gas may be oxidized completely. A temperature of 1100–1200°C may be achieved with this burner. It is used for ignition of the precipitates and inverting them into weighable forms.

Hot Plates

The electrically heated hot plate (with magnetic stirrer) preferably provided with temperature control is of great importance in the analytical laboratory. These are used for different heating operations with continuous stirring in the laboratory. The stirring is accomplished with the help of small permanent magnets sealed in a teflon.

Electric Ovens

The most convenient type is an electrically heated thermostatically controlled drying oven having a temperature range from room temperature to about 250–300°C, the temperature can be controlled to within ± 1–2°C. They are used for drying precipitates, solids, glassware etc. These are also available with digital pyrometer display unit (Fig. 2.7).

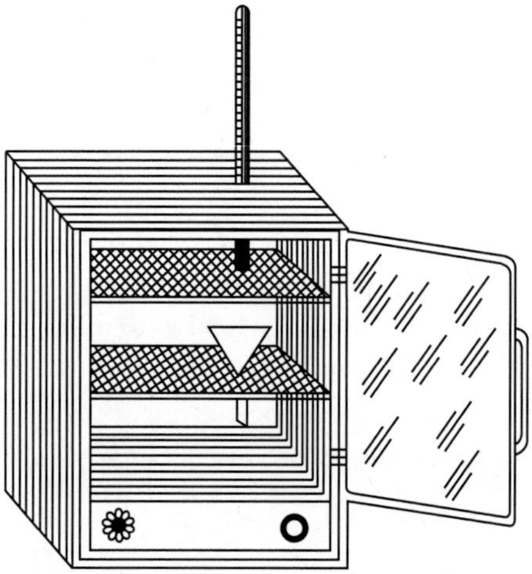

FIGURE 2.7 **Electric oven.**

Muffle Furnace

An electrically heated furnace of muffle form should be available in every well-equipped laboratory. The maximum temperature of 1200°C may be achieved by muffle furnaces. These are generally used for gravimetric estimations, fuel analysis etc. (Fig. 2.8). Many other glasswares and equipments being used in Laboratories in common practice are shown in Fig. 2.9.

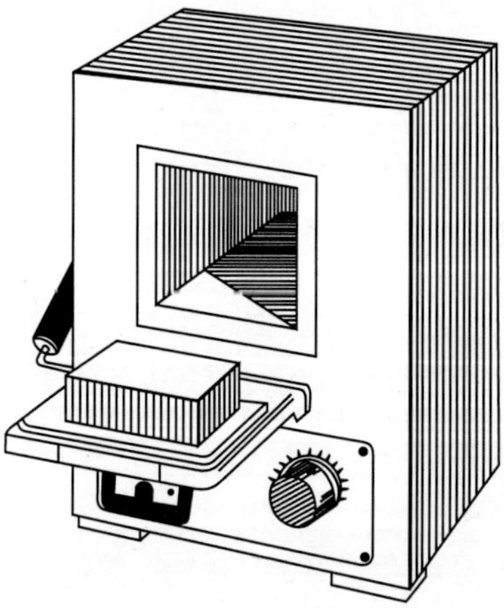

FIGURE 2.8 **Muffle furnace.**

2.3 COMMON LABORATORY APPARATUS

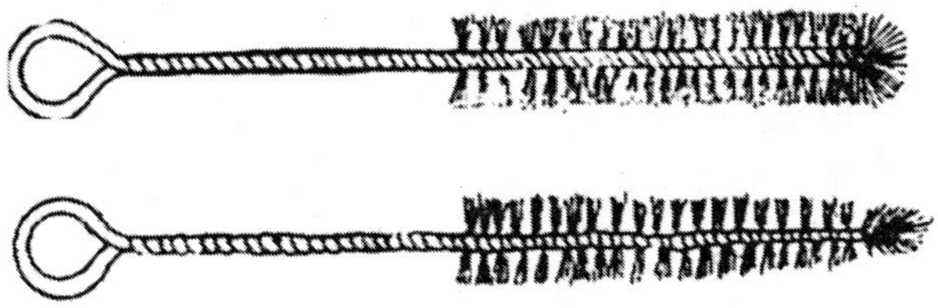

(*a*) Cleaning brushes.

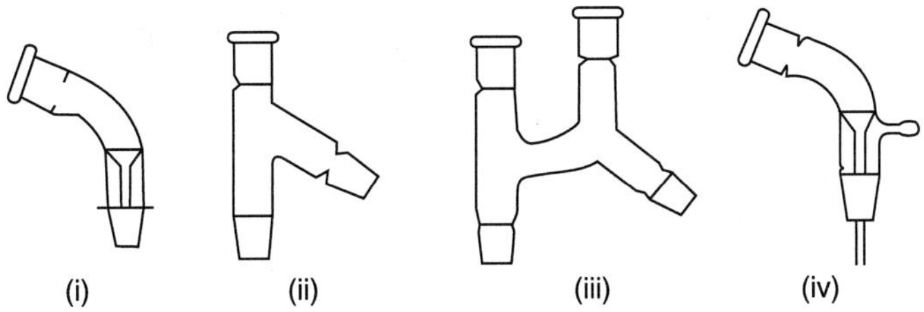

(*b*) Adapters.

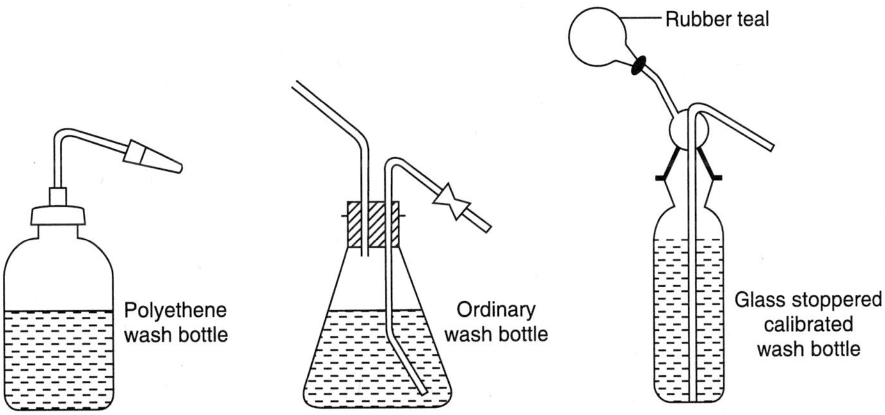

(*c*) Wash bottle.

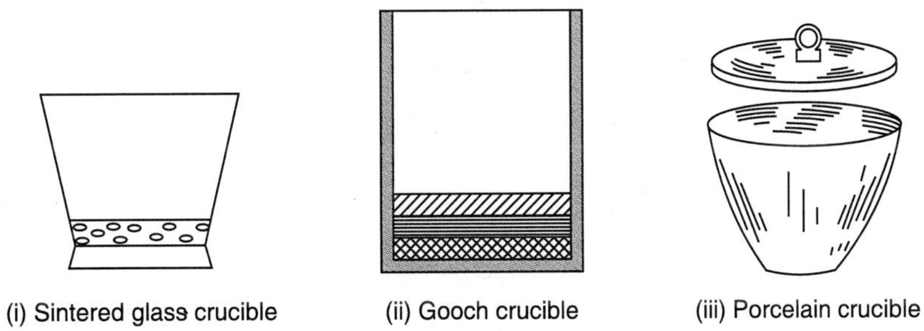

(*d*) Crucible.

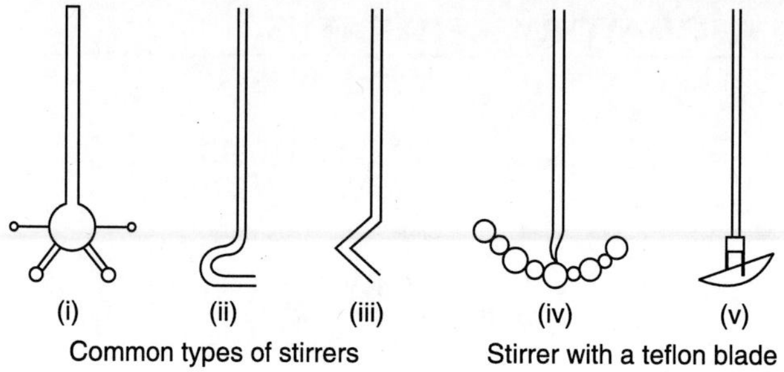

(*e*) Stirrer.

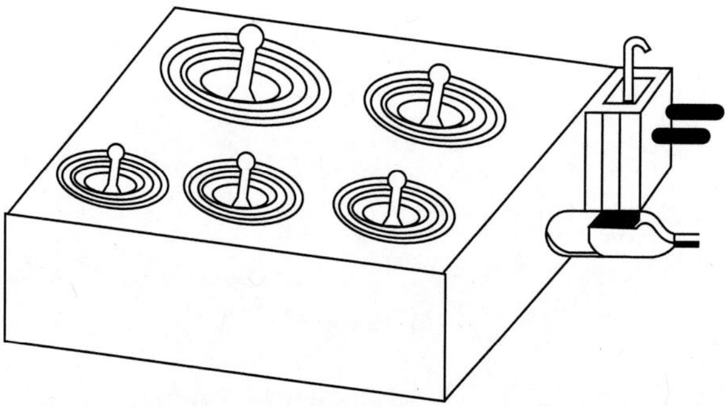

(*f*) Water bath (electrically heated).

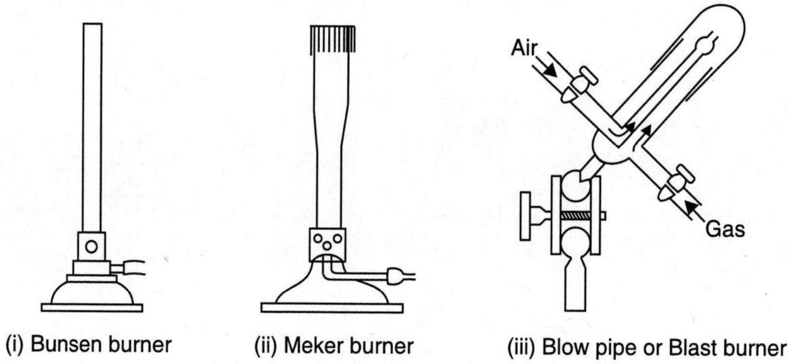

(*g*) Burner.

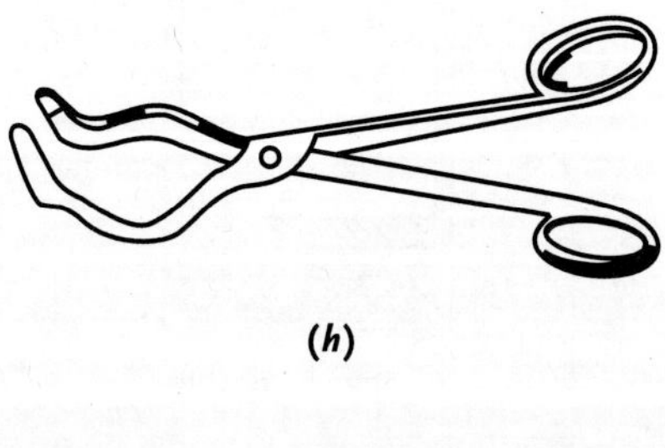

(*h*)

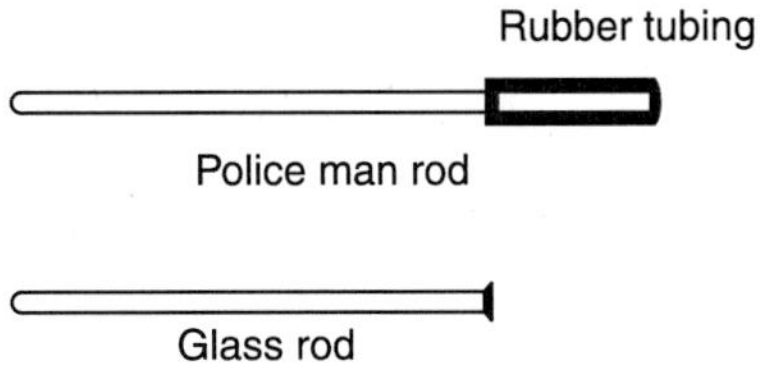

(*i*) **Glass rod.**

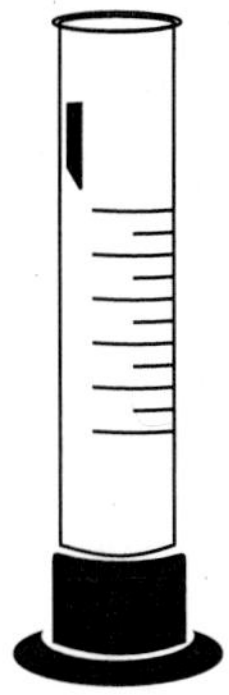

(*j*) **Graduated measuring cylinder.**

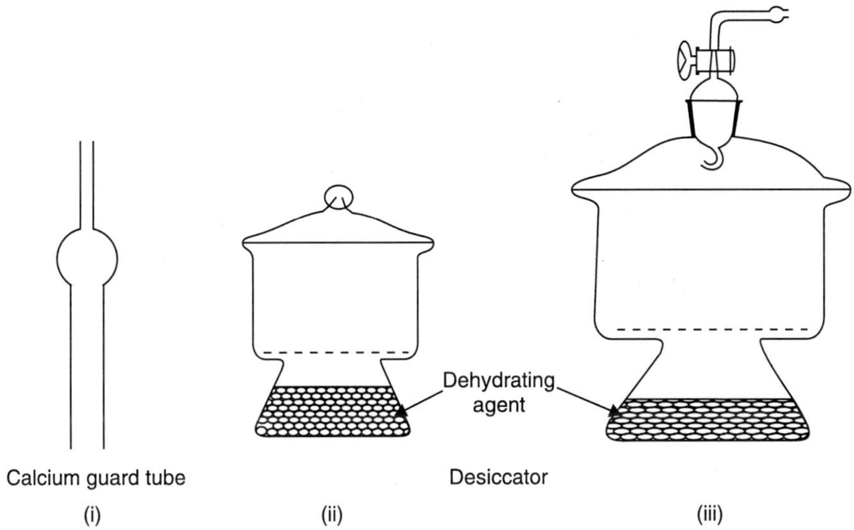

(*k*) **Drying apparatus.**

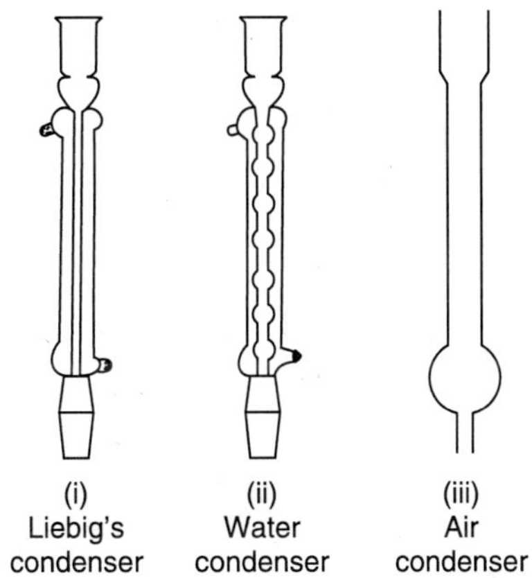

(*l*) **Condensers.**

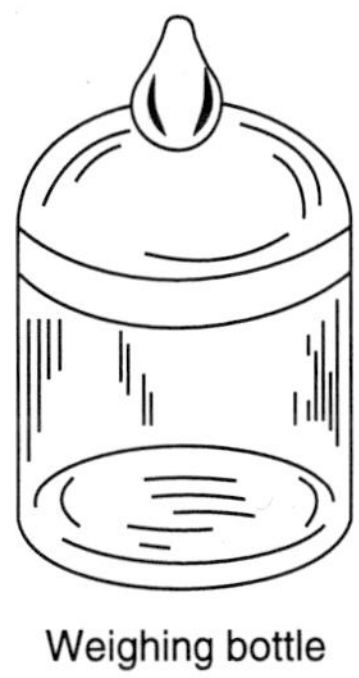

(*m*) **Weighing bottle.**

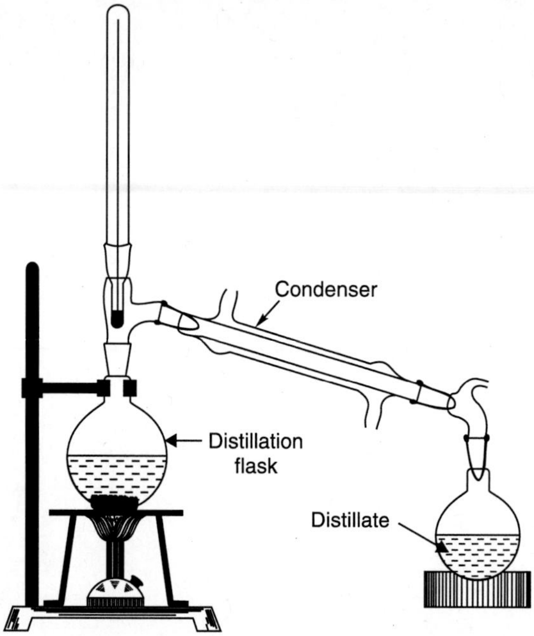

(*n*) Apparatus for simple distillation.

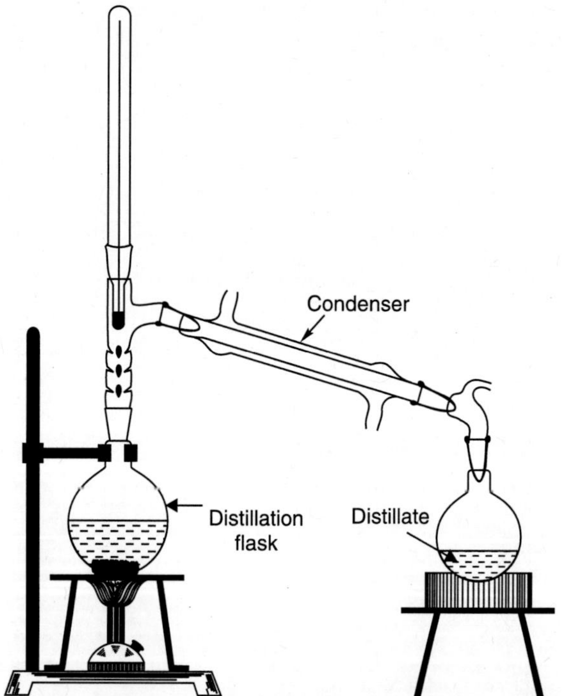

(*o*) Apparatus for fractional distillation.

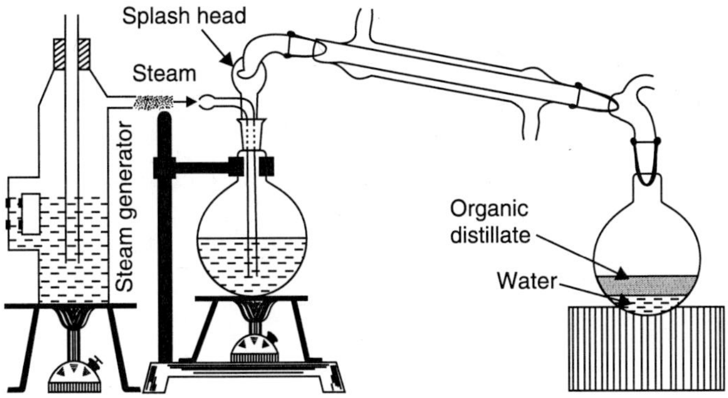

(*p*) Apparatus for steam distillation.

FIGURE 2.9 Common laboratory apparatus.

Chapter 3 TITRIMETRY AND GRAVIMETRY

3.1 INTRODUCTION

Chemical analysis is carried out in two ways *i.e.,* **qualitative anlaysis** and quantitative analysis. Qualitative analysis is the detection of species present in an unknown sample. Quantitative analysis involves the determination of quantity of a particular component present in the substance. Quantitative analysis can be carried out in two ways.

(*i*) *Gravimetric Analysis:* When the analysis is based on the measurement of weight then it is called gravimetric analysis.

(*ii*) *Volumetric Analysis:* When the analysis is based on measuring the volume of the solution of substance then it is called volumetric analysis.

3.2 TITRIMETRIC (VOLUMETRIC) ANALYSIS

Titrimetric or volumetric analysis is a technique of finding the volume of one solution which will exactly react with a definite volume of another solution of known concentration. In this method a known volume of one solution is taken in any glass vessel (conical flask) and the solution of other substance is added very slowly to it till the reaction is complete. This process is known as titration.

From the volume of unknown solution which is required to completely react with definite volume of the other solution of known strength, the strength of the unknown solution can be calculated using the Law of Equivalence or Normality Equation.

Normality Equation

According to the law of equivalence, the substances react with each other in a definite ratio by mass. For example, if the solution of a substance 1 is titrated against the solution of a substance 2 then at the end point according to normality equation.

Number of gram equivalents of solution 1 = Number of gram equivalents of solution 2

or, $$N_1V_1 = N_2V_2.$$

where N_1 and N_2 are the normalities of two solutions 1 and 2 and V_1 and V_2 are the volumes of two solutions 1 and 2 respectively. This is known as the normality equation and is the fundamental principle of all volumetric calculations.

If N_1, V_1 and V_2 are known, then unknown normality of a solution can be calculated,

$$N_2 = \frac{N_1V_1}{V_2} = x \text{ (say)}$$

After knowing the normality of a solution, its strength in grams per litre can be found out by applying the following relation:

$$\begin{aligned}\text{Strength} &= \text{Normality} \times \text{Equivalent weight}\\ &= x \times \text{Equivalent weight.}\end{aligned}$$

Terms Used In Volumetric Analysis

1. **Titration:** The process of finding out the volume of one of the solution, required to react completely with the definite volume of the other solution is known as titration.
2. **Titrant:** The solution of known strength is called the titrant.
3. **Titrate:** The solution which contains a substance to be estimated or whose strength is to be estimated is called as titrate.
4. **End Point:** The point at which the completion of reaction occurs is called equivalence point, whereas an end point is an observable physical change that occurs near the equivalence point. The equivalence point determined using an indicator is called the end point. Since, the colour change of the indicator generally occurs only after the equivalence point and hence end point differs slightly from equivalence point. Equivalence point can be determined by instruments (pH meter, conductivity meter etc.)
5. **Indicator:** The substance which indicates the end point of titration by change in colour is called an indicator.
6. **Strength:** The amount of substance dissolved in one litre of a solution when expressed in grams is termed as strength of a solution. Strength of the solution can also be expressed in any of the following ways:

   (*i*) *Normality (N):* It is the number of gram equivalents of the substance dissolved per litre of the solution. It is denoted by N.

$$N = \frac{\text{Grams of solute per litre of the solution}}{\text{Gram equivalent weight of the solute}}.$$

Normal Solution: It contains 1 gm equivalent of solute per litre of solution. A solution containing 0.1 gm equivalent of solute per litre of solution, is called deci-normal solution and is denoted by N/10.

(*ii*) *Molarity (M):* It is the number of moles of the solute present per litre of the solution. It is denoted by M.

$$M = \frac{\text{Grams of solute per litre of the solution}}{\text{Molecular mass of the solute}}.$$

(*iii*) *Molality (m):* It is the number of moles of the substance dissolved in 1000 gm or 1 kg of the solvent. It is denoted by *m*.

$$m = \frac{\text{Grams of solute per 1000 g of the solvent}}{\text{Molecular mass of the solute}}.$$

(*iv*) *Formality (F):* It is the number of gram formula weight of the substance dissolved in one litre of the solution. It is denoted by F.

$$F = \frac{\text{Grams of solute per litre of the solution}}{\text{Formula weight}}.$$

7. **Standard Solutions:** A solution of known concentration is known as standard solution. There are various substances used for making standard solutions in volumetric analysis being classified into following two types,

(*i*) *Primary Standard:* The substance whose standard solution can be prepared directly by dissolving an accurately weighed amount of it in a definite volume of solvent or solution is known as a primary standard.

e.g., anhydrous sodium carbonates, oxalic acid, succinic acid, $AgNO_3$, $K_2Cr_2O_7$, dried KCl, $FeSO_4$ $(NH_4)_2SO_4 \cdot 6H_2O$ etc. A primary standard should possess following characteristics.

(*a*) It should be easily available in pure and dry form.

(*b*) It should have high equivalent weight so that error due to weighing is minimised.

(*c*) It must not undergo decomposition in presence of solvent.

(*d*) It should be readily soluble in water or the desired solvent.

(*e*) It should neither be hygroscopic nor efflorescent.

(*f*) Its composition must not change on standing.

(*g*) It should be stable and unaffected by air.

(*ii*) *Secondary Standard:* The substances which do not fulfil the above requirements and their standard solutions are not prepared directly by weighing are called secondary standard substances. The solution of this type of substance firstly prepared is of approximate strength by weighing. Then the exact strength of the solution is found by titrating it against some primary standard. This process is called **standardisation** *e.g.*, $KMnO_4$, inorganic acids, alkali hydroxides etc.

Classification of Methods of Volumetric Analysis

Volumetric analysis can be classified into following four types depending on the reactions used in volumetric analysis.

I. Neutralization Titrations or Acid Base Titrations: Acid base titrations are the titrations in which acid is titrated against an alkali or vice versa. The reaction involves the combination of hydroxyl ions with the hydrogen ions to form water. A typical example is the titration of sodium hydroxide against hydrochloric acid.

$$NaOH + HCl \longrightarrow NaCl + H_2O$$

or $$Na^+ + OH^- + H^+ + Cl^- \longrightarrow Na^+ + Cl^- + H_2O$$

or $$H^+ + OH^- \longrightarrow H_2O$$

Acid base titrations are also known as acidimetry and alkalimetry.

Acidimetry: To find the strength of an unknown acid with the help of a known base is called acidimetry.

Alkalimetry: To find the strength of an unknown base with the help of a known acid is called alkalimetry. The equivalence point in neutralization reactions occurs at a definite pH value. The substances which change colour with pH are used to indicate the end point of acid base titrations. Acid-base indicators are complex organic compounds which have one colour in acid solution and different colour in alkaline solution.

Theory of Acid Base Indicators

1. **Ostwald's Theory:** According to this theory the indicators themselves are weak acids or weak bases. Action of phenolphthalein and methyl orange on the basis of this theory is explained.

Action of Phenolphthalein: Phenolphthalein is a weak acid and can be represented as HPh. Its unionized molecules are colourless while on ionization it gives pink coloured Ph^- ions.

$$\underset{\substack{\text{(Colourless)}\\\text{Acid colour}}}{HPh} \rightleftharpoons \underset{\text{(Colourless)}}{H^+} + \underset{\substack{\text{(Pink)}\\\text{Base colour}}}{Ph^-}$$

In presence of acid, due to increase in concentration of common H^+ ions in solution the dissociation of HPh is suppressed and thus the solution becomes colourless on the other hand, when a base like NaOH is added, the OH^- ions of the base combine with H^+ of the indicator to give unionized water. This results shifting of equilibrium towards right, producing more of Ph^- ions. As such, the solution appears pink.

Action of Methyl Orange: Methyl orange is a weak base and can be represented as Me. It changes colour from yellow to red depending upon the pH of the solution. The equilibrium can be represented as.

$$\underset{\substack{\text{(Yellow)}\\\text{Base colour}}}{Me} + H_2O \rightleftharpoons \underset{\substack{\text{(Red)}\\\text{Acid colour}}}{MeH^+} + OH^-$$

In presence of acid, the hydrogen ions imparted by acid combine with OH^- ions and the equilibrium shifts to right and hence the solution appears red. On the other hand, if a base is added because of common ion effect equilibrium shifts to left and hence the solution appears yellow.

2. **Modern Quinoid Theory:** According to this theory, the acid-base indicators can exist in two tautomeric forms *i.e.*, benzenoid and quinoid form. The two forms have different colours. Change in pH causes the conversion of benzenoid form to quinoid form or vice versa and hence change in colour.

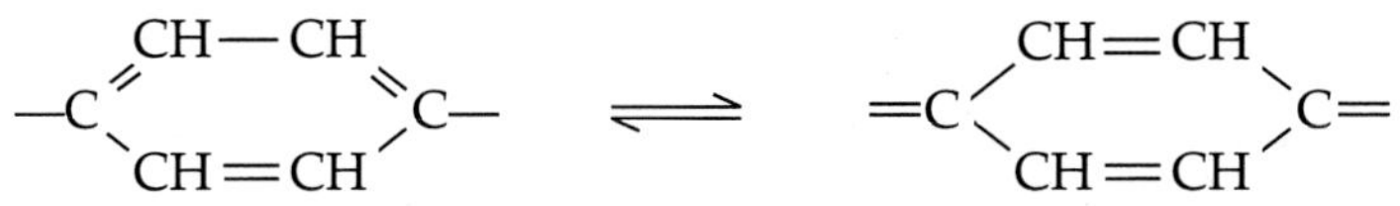

(Benzenoid form) (Quinoid form)

The action of the phenolphthalein and methyl orange indicators according to this theory is explained as follows.

Action of Phenolphthalein: Phenolphthalein possesses a benzenoid structure in acidic medium and it changes to pink coloured quinoid form in alkaline medium.

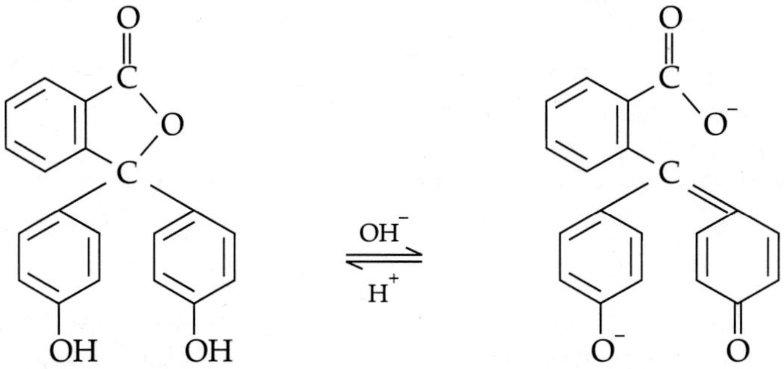

Colourless benzenoid form (Acid medium)

Red colour quinoid form (Alkaline medium)

Action of Methyl Orange: Methyl orange possesses a benzenoid form in alkaline solution and is yellow in colour. In acidic solution it exists in red colour having quinoid structure.

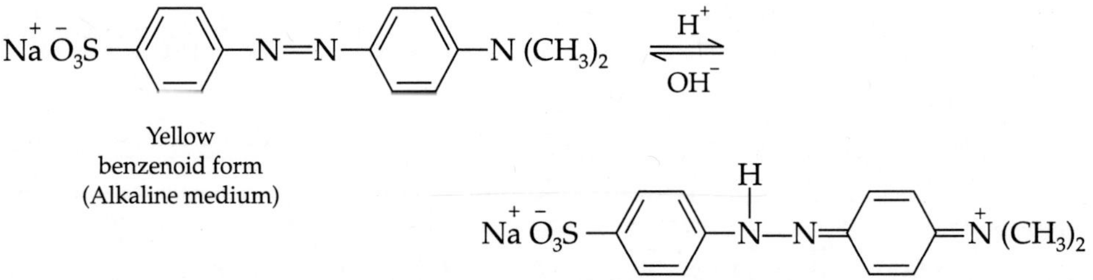

Yellow benzenoid form (Alkaline medium)

Red quinoid form (Acid medium)

Selection of Indicator: Different indicators have different pH ranges. Therefore it is important to select a proper indicator for a given titration. Following are the general principles for selecting a suitable indicator (Table 3.1).

TABLE 3.1

Solution		*Indicator*
Acid	*Base*	
Strong	Strong	Any
Strong	Weak	Weak base or strong acid *e.g.*, methyl orange, *p*-nitrophenol
Weak	Strong	Weak acid *e.g.*, phenolphthalein, litmus
Weak	Weak	None satisfactory

II. Redox Titrations: This involves the change in the oxidation number or transfer of electrons among the reactants and products. In a redox titration, a reducing agent is titrated against an oxidising agent or vice versa.

e.g., Potassium permanganate as the oxidising agent is titrated against oxalic acid as reducing agent.

Redox Indicators: A redox indicator is defined as a substance whose oxidised form is of different colour form that of its reduced form. The change in colour of redox indicator is dependent upon the electrode potential of the system.

Redox indicators can be classified as:

(*i*) *Self Indicator:* If the reacting solution is strongly coloured and is decolourized or changes into a different colour by the substance that is being determined then it is called self indicator.

For example potassium permanganate is purple in its oxidized form, but its reduced product *i.e.*, Manganese (II) ion is almost colourless.

$$\underset{\text{Purple}}{MnO_4^- + 8H^+ + 5e^-} \longrightarrow \underset{\text{Colourless}}{Mn^{2+} + 4H_2O}$$

Therefore during the titration of $KMnO_4$ with a reducing agent such as oxalic acid, Manganese (II) is produced which is colourless. As the titration proceeds, after the whole of reducing agent being consumed up, addition of an excess drop of $KMnO_4$ solution gives its own colour (pink) to the solution. This is the end point of the reaction. Hence $KMnO_4$ acts as self indicator since no other indicator is added to mark the end point.

(*ii*) *Internal indicator:* Internal indicator is the substance which is added to the volumetric flask in which titration is carried out.

For exmaple, acid base indicators like phenolphthalein, methyl orange etc.

(*iii*) *External Indicator:* The indicator which is used externally is called external indicator. This can not be added to the volumetric flask in which the titration is carried out.

For example, potassium ferricyanide is used as an external indicator in the titration of $K_2Cr_2O_7$ with ferrous ammonium sulphate. Few drops of this indicator are taken on the glazed white tile and a drop of the oxidised solution is taken out with a glass rod out of titration flask and touched with the drop of indicator. Titration is continued till the reaction mixture gives no blue colour.

Iodine Titration: Iodine is a mild oxidising agent and is used in volumetric estimations. The redox titration using iodine directly or indirectly as an oxidising agent are called **Iodine Titrations**. These are of two types.

1. *Iodimetric Titrations (Iodimetry):* Iodimetry titrations are defined as those iodine titrations in which a standard iodine solution is used as an oxidant and iodine is directly titrated with a reducing agent. Iodimetry is employed to determine the strength of reducing agents such as sodium thiosulphate, sulphite, arsenite, stannous and ferrous ions in neutral or moderately acidic solutions.

$$2Na_2S_2O_3 + I_2 \longrightarrow Na_2S_4O_6 + 2NaI$$

2. *Iodometric Titrations (Iodometry):* Iodometric titrations are defined as those iodine titrations in which some oxidising agent liberates iodine from iodide and then liberated iodine is titrated against standard solution of reducing agent added from a burette. Iodometry is used to estimate the strength of oxidising agents.

For example, the estimation of Cu^{2+} with thiosulphate.

$$2Cu^{2+} + 4I^- \longrightarrow Cu_2I_2 + 2I_2$$

$$I_2 + 2S_2O_3^{2-} \longrightarrow S_4O_6^{2-} + 2I^-$$

The indicator used in iodine titrations is starch solution which is used as an internal indicator. Starch forms a dark blue complex with triiodate ions. The blue colour disappears when there is no more free iodine.

III. Precipitation Titration (Argentometric Titration): Precipitation titration results in the formation of a precipitate. An example for this type is the titration of silver nitrate against sodium chloride.

$$AgNO_3 + NaCl \longrightarrow AgCl\downarrow + NaNO_3$$

The titrations accompanied by the formation of sparingly soluble salts, when the solutions of two reacting substances are brought into contact with each other, are called precipitation titrations. Since the reactions are limited to those involving $AgNO_3$ solution and hence are called **Argentometric titrations.** The use of indicators in precipitation titration is based on the formation of coloured precipitates near the end point.

IV. Complexometric Titration: Complexometric titration is a generic term involving the titrations based on the formation of a soluble complex between a metal ion and a complexing agent.

Sodium salt of EDTA (ethylene diamene tetraacetic acid) is often used as a complexing reagent in titrations. The indicators used in these titrations are the organic colouring substances that undergo change of colour when they form metal complexes. The indicator forms a less stable complex with the metal atom than EDTA-metal complex. The indicator releases metal ion at equivalence point and this results into change in colour. Some of indicators include, Eriochrome Black T, Calcon, Zincon etc.

Following are the two types of complexometric titrations:

(*a*) *Direct Titration:* The solution containing the metal ion to be detected is buffered to the desired pH and titrated directly with the standard EDTA solution.

(*b*) *Back Titration:* Many metals can not be titrated directly, thus they may precipitate from the solution in the pH range necessary for the titration, or they may form inert complexes, or a suitable metal indicator is not available. In this case, an excess of standard EDTA solution is added, the resulting solution is buffered to the desired pH and the excess of the EDTA is back titrated with a standard metal ion solution ($ZnCl_2$ or $ZnSO_4/MgCl_2$ or $MgSO_4$). The end point is detected with the help of the metal indicator which responds to the zinc or magnesium ions introduced in the back titration.

EXPERIMENT 3.1

Aim of the Experiment

Standardisation of $KMnO_4$ solution.

Apparatus Requires

- Burette
- Pipette
- Conical flask
- Glazed tile.

Chemicals Required

- Given $KMnO_4$ solution
- 5 N sulphuric acid solution
- $\frac{N}{10}$ oxalic acid solution.

Theory

$KMnO_4$, as it is not available in pure form and is associated with manganese dioxide which catalyse the auto decomposition of permanganate solution on standing. Because of this concentration of $KMnO_4$ solution falls somewhat after preparation. Hence standard $KMnO_4$ solution cannot be prepared by exact weighing. It has to be standardised by oxalic acid.

$$2KMnO_4 + 3\,H_2SO_4 \longrightarrow K_2SO_4 + 2MnSO_4 + 5[O] + 3H_2O$$

$$5C_2H_2O_4 + 5[O] \longrightarrow 5H_2O + 10CO_2$$

$$2KMnO_4 + 3\,H_2SO_4 + 5C_2H_2O_4 \longrightarrow K_2SO_4 + 2MnSO_4 + 8H_2O + 10CO_2$$

$KMnO_4$ acts as a self indicator and appearance of light pink colour which remains for 1–2 min. is the end point.

Procedure

Pipette out 10 ml of standard oxalic acid solution into a conical flask. To this add 10 ml of 5N H_2SO_4. Warm the solution over a water bath upto 50–60°C temperature. Now titrate it against $KMnO_4$ solution filled in burette until pale pink colour persists for 1 minute. Note down the volume of $KMnO_4$ used (V_2 ml) Repeat the titration to get at least two concordant readings.

Tabulation

No. of observations	*Volume of oxalic acid solution (V_1)*	*I.B.R.*	*F.B.R.*	*Volume of $KMnO_4$ used (V_2) Difference in ml*
1	10			
2	10			
3	10			

Calculation

$$N_1 = \text{Normality of oxalic acid} = \frac{N}{10}$$

$$V_1 = \text{Volume of oxalic acid} = 10 \text{ ml}$$

$$N_2 = \text{Normality of } KMnO_4 = ?$$

$$V_2 = \text{Volume of } KMnO_4 = x \text{ ml (say)}$$

Applying normality equation

$$N_1V_1 = N_2V_2$$

or,

$$N_2 = \frac{N_1V_1}{V_2} = \frac{\frac{N}{10} \times 10}{x} = \frac{N}{x}.$$

Conclusion

The normality of $KMnO_4$ solution is $\frac{N}{x}$.

Preparation of Reagents

1. *Oxalic Acid (N/10):* Dissolve 6.3 gm of oxalic acid (*e.q.* wt. 63 gm) in distilled water and dilute to 1 litre.
2. *Sulphuric Acid (5 N):* Add slowly 139 ml of conc. sulphuric acid (36 N, 98%) to distilled water. After cooling make up the volume to 1 litre with distilled water.

EXPERIMENT 3.2

Aim of the Experiment

Standardisation of hypo (sodium thiosulphate) solution.

Apparatus Required

- Burette
- Pipette
- Stoppered conical flask
- Glazed tile.

Chemical Required

- Given sodium thiosulphate solution
- Potassium iodide
- Sodium bicarbonate
- $\frac{N}{20}$ $K_2Cr_2O_7$ solution.
- Starch indicator.

Theory

Sodium thiosulphate ($Na_2S_2O_3$, $5H_2O$) is readily obtainable in a state of high purity, but there is always some uncertainty as to the exact water content because of efflorescent nature of the salt and for some other reasons. The substance is therefore unsuitable as a primary standard. It is a reducing agent by virtue of the reaction.

$$2S_2O_3^{2-} \rightleftharpoons S_4O_6^{2-} + 2e^-$$

An approximately 0.1 M solution is prepared by dissolving about 25 g crystallised sodium thiosulphate in 1L of water in a graduated flask. The solution is standardised with potassium dichromate. Potassium dichromate is reduced by an acid solution of potassium iodide and iodine is set free.

$$Cr_2O_7^{2-} + 6I^- + 14H^+ \longrightarrow 2Cr^{3+} + 3I_2 + 7H_2O$$

Procedure

Transfer 100 ml of cold freshly boiled distilled water in a 250 ml conical flask, add about 2 g of KI and 2 g of $NaHCO_3$ and shake until the salts are dissolved. Add 6 ml of conc. HCl slowly by gently shaking the flask. Add 25 ml of $\frac{N}{20}$ $K_2Cr_2O_7$ solution and mix the solution well. Wash the inner sides of the flask with a little distilled water and stopper the flask. Allow to stand for 5 minutes for completion of the reaciton. Rinse the stopper with distilled water and titrate the liberated iodine with the hypo solution from the burette. When most of the iodine has reacted, the solution acquires greenish yellow colour. At this stage add 1 ml of starch solution when the solution turns blue due to the formation of starch iodide complex. Continue the titration until the greenish blue colour changes to light green by addition of a single drop of hypo at the end point. Note down the volume of hypo used (V_2 ml) Repeat the process to get the concordant volume of hypo used.

Tabulation

No. of observations	*Volume of $K_2Cr_2O_7$ solution (V_1 ml)*	*I.B.R.*	*F.B.R.*	*Volume of $Na_2S_2O_3$ solution used (Difference V_2 ml)*
1	25			
2	25			
3	25			

Calculation

$$N_1 = \text{Normality of } K_2Cr_2O_7 = \frac{N}{20}$$

$$V_1 = \text{Volume of } K_2Cr_2O_7 = 25 \text{ ml}$$

$$N_2 = \text{Normality of hypo} = ?$$

$$V_2 = \text{Volume of hypo} = x \text{ ml (say)}$$

Applying normality equation,

$$N_1V_1 = N_2V_2$$

or,
$$N_2 = \frac{N_1V_1}{V_2} = \frac{\frac{N}{20} \times 25}{x} = \frac{1.25}{x} \text{ N.}$$

Conclusion

The normality of sodium thiosulphate solution is $\frac{1.25}{x}$ N.

Preparation of Reagents

1. *Potassium Dichromate (N/20):* Dissolve 2.451 g of pure dry potassium dichromate in distilled water and dilute to 1L.
2. *Starch Indicator (1%):* Prepare a paste of 1g of powdered starch with distilled water and add to this paste 100 ml of boiled water with constant stirring. Boil for about five minutes and then cool.

EXPERIMENT 3.3

Aim of the Experiment

Standardisation of hydrochloric acid solution.

Apparatus Required

- Burette
- Pipette
- Conical flask
- Glazed tile.

Chemicals Required

- Given hydrochloric acid solution
- 0.1 N Na_2CO_3 solution
- Methyl orange indicator.

Theory

Hydrochloric acid is a secondary standard because its concentration is known only approximately. Therefore it needs to be standardised before it is used. It has to be standardised by sodium carbonate, using methyl orange as indicator to know the exact normality of HCl solution.

$$CO_3^{2-} + 2H^+ \longrightarrow H_2O + CO_2$$

Procedure

Pipette out 20 ml of 0.1 N Na_2CO_3 solution into a 100 ml conical flask. Add 2–3 drops of methyl orange indicator, the solution would turn yellow on addition of this indicator. Titrate it with the given hydrochloric acid solution taken in a burette. Continue the titration with shaking till a permanent red colour is obtained as the end point. Note down the volume of HCl used (V_2 ml). Repeat the titration to get concordant readings.

Tabulation

No. of observations	*Volume of Na_2CO_3 solution (V_1 ml)*	*I.B.R. (ml)*	*F.B.R. (ml)*	*Volume of HCl solution used (Difference V_2 ml)*
1	20			
2	20			
3	20			

Calculation

$$N_1 = \text{Normality of } Na_2CO_3 = \frac{N}{10}$$

$$V_1 = \text{Volume of } Na_2CO_3 = 20 \text{ ml}$$

$$N_2 = \text{Normality of HCl} = ?$$

$$V_2 = \text{Volume of HCl} = x \text{ ml (say)}$$

Applying normality equation,

$$N_1V_1 = N_2V_2$$

or,

$$N_2 = \frac{N_1V_1}{V_2} = \frac{\frac{N}{10} \times 20}{x} = \frac{2N}{x}.$$

Conclusion

The normality of hydrochloric acid solution is $\frac{2}{x}$ N.

Preparation of Reagents

1. *Sodium Carbonate (0.1 N):* Dissolve 5.30 g of Na_2CO_3 dried at 250°C in boiled out distilled water and dilute to 1 litre with distilled water.

2. *Methyl Orange Indicator:* Dissolve 50 mg of free acid form of methyl orange solid in distilled water and make the volume to 100 ml with distilled water. Filter the solution if necessary.

EXPERIMENT 3.4

Aim of the Experiment

Determination of Fe^{2+} ion in Mohr's salt.

Apparatus Required

- Burette
- Pipette
- Conical flask
- Glazed tile.

Chemicals Required

- $\frac{N}{10}$ $KMnO_4$ solution
- 5 N sulphuric acid solution
- Given Mohr's salt solution.

Theory

Potassium permanganate oxidises ferrous ion into ferric ion in acid medium in cold and is itself reduced to colourless manganous ions. The estimation is based on the oxidation reduction reaction given below.

$$MnO_4^- + 8H^+ + 5e^- \longrightarrow Mn^{2+} + 4H_2O$$

$$[Fe^{2+} \longrightarrow Fe^{3+} + e^-] \times 5$$

$$MnO_4^- + 5\,Fe^{2+} + 8H^+ \longrightarrow Mn^{2+} + 5\,Fe^{3+} + 4H_2O$$

$$2KMnO_4 + 10\,FeSO_4 + 8\,H_2SO_4 \longrightarrow 2MnSO_4 + 5Fe_2(SO_4)_3 + K_2SO_4 + 8H_2O$$

Procedure

Pipette out 10 ml of the Mohr's salt solution into the conical flask. Add 10 ml of 5N H_2SO_4 solution into this. Titrate it against $\frac{N}{10}$ $KMnO_4$ solution filled in burette till a pale pink colour persists for at least 1–2 minutes. Note down the volume of $KMnO_4$ used (V_2 ml). Repeat the titration to get at least two concordant readings.

Tabulation

No. of observations	*Volume of Mohr's salt solution (V_1 ml)*	*I.B.R. (ml)*	*F.B.R. (ml)*	*Volume of $KMnO_4$ solution used (Difference V_2 ml)*
1	10			
2	10			
3	10			

Calculation

N_1 = Normality of Mohr's salt solution = ?

V_1 = Volume of Mohr's salt solution = 10 ml

N_2 = Normality of $KMnO_4 = \frac{N}{10}$

V_2 = Volume of $KMnO_4 = x$ ml (say)

Applying normality equation,

$$N_1V_1 = N_2V_2$$

or,

$$N_1 = \frac{N_2V_2}{V_1} = \frac{\frac{N}{10} \times x}{10} = \frac{x}{100}\ \text{N}$$

Strength of Fe^{2+} ions in gm/litre = Eq. wt. × normality

$$= 56 \times \frac{x}{100}$$

$$= 0.56 \times x \text{ gm/litre.}$$

Conclusion

The given Mohr's salt solution contains $0.56 \times x$ gm/litre of Fe^{2+} ion.

Preparation of Reagents

1. *Potassium Permanganate (N/10):* Dissolve 3.1607 g of $KMnO_4$ (Eq. wt. 31.607 g) in distilled water. Boil for 1 hour, cool, filter and make up the volume to 1 litre with distilled water. Standardise it with oxalic acid to know the exact strength.
2. *Sulphuric Acid (5 N):* Add slowly 139 ml of conc. sulphuric acid (36 N, 98%) to distilled water. After cooling make up the volume to 1 litre with distilled water.

N.B. The solution of ferrous ammonium salt (Mohr's salt) is prepared in dil. H_2SO_4 in order to avoid oxidation and hydrolysis. To prepare Mohr's salt solution weigh 20 g of hydrated Mohr's salt, add 20 ml of dil. H_2SO_4 and distilled water and dissolve the salt. Make up the volume to 1 litre.

EXPERIMENT 3.5

Aim of the Experiment

Determination of ferrous (Fe^{2+}) and ferric (Fe^{3+}) ions in the given mixture.

Apparatus Required

- Burette
- Pipette
- Conical flask
- Glazed tile.

Chemicals Required

- Given iron solution
- $\frac{N}{10}$ $KMnO_4$ solution
- 5 N H_2SO_4 solution
- Concentrated hydrochloric acid
- Concentrated solution of stannous chloride
- Mercuric chloride solution (5%).

Theory

The amount of ferrous ions present in the given mixture solution is determined by titrating an aliquot of the solution with a standard solution of $KMnO_4$. Ferric salts are not oxidised by permanganate. The ferric ions present in the solution are reduced to ferrous state and the titration against $KMnO_4$ estimates the total iron (Fe^{2+} and Fe^{3+}) present in the solution. The difference in the two titre values corresponds to the ferric ion present in the solution. The reduction is commonly effected by the action of stannous chloride.

$$2\,FeCl_3 + SnCl_2 \longrightarrow 2FeCl_2 + SnCl_4$$

The excess $SnCl_2$ must be removed completely as it is also oxidised by $KMnO_4$. This is done by means of mercuric chloride.

$$SnCl_2 + 2HgCl_2 \longrightarrow SnCl_4 + Hg_2Cl_2$$

Mercurous chloride (Hg_2Cl_2) is deposited in the form of silky white precipitate. This is not filtered off and the liquid is titrated with $KMnO_4$.

$$Fe^{3+} \xrightarrow[+e^-]{SnCl_2} Fe^{2+} \xrightarrow[-e^-]{KMnO_4} Fe^{3+}$$

Fe^{3+} initially present is reduced to an equivalent amount of Fe^{2+} which is subsequently oxidised to Fe^{3+}. Since, only one electron per atom is involved in each of these steps.

Equivalent wt. of ferric salt = equivalent wt. of ferrous salt ≡ atomic wt. of Fe.

Procedure

Pipette out 10 ml of the given iron solution and add 10 ml 5 N H_2SO_4 solution in it and titrate it against $\frac{N}{10}$ $KMnO_4$ solution (standardised) taken in the burette. The pale pink colour shows the end point. Repeat the experiment to get at least two concordant readings. Note down the volume of $KMnO_4$ used (V_1 ml) which corresponds to ferrous iron present in the solution.

Again pipette out 10 ml of given iron solution in a conical flask and add 10 ml concentrated hydrochloric acid into it. Boil the contents in the conical flask for about 10 minutes. Remove the burner and add carefully, drop by drop with continuous stirring $SnCl_2$ solution until the yellow colour of the solution has nearly disappeared. Add a drop of the dilute $SnCl_2$ in excess to ensure that the reduction is complete.

Cool the solution, dilute it with 100 ml in distilled water, add 20 ml of $HgCl_2$ solution, stir thoroughly and leave to stand for about 2 minutes. A slight silky white precipitate of mercurous chloride should be obtained. Without filtering off the precipitate add 10 ml of 5 N H_2SO_4 solution and titrate it with standard $KMnO_4$ solution taken in the burette. A pale pink colour marks the end point. Repeat the experiment to get at least two concordant readings. Note down the volume of $KMnO_4$ used (N_2 ml).

Tabulation

Ferrous Iron

No. of observations	*Volume of iron solution (ml)*	*I.B.R. (ml)*	*F.B.R. (ml)*	*Volume of $KMnO_4$ solution used (Difference V_1 ml)*
1	10			
2	10			
3	10			

Total Iron

No. of observations	*Volume of iron solution (ml)*	*I.B.R. (ml)*	*F.B.R. (ml)*	*Volume of $KMnO_4$ solution used (Difference V_2 ml)*
1	10			
2	10			
3	10			

Calculation

Ferrous Iron: We know that

$$1000 \text{ ml of 1N } KMnO_4 \equiv 55.85 \text{ g of } Fe^{2+}$$

$$\therefore \quad V_1 \text{ ml of } \frac{N}{10} KMnO_4 \equiv \frac{55.85}{1000} \times \frac{1}{10} \times V_1 gFe^{2+}$$

This is the amount of ferrous iron present in 10 ml of the given solution

$$\therefore \quad \text{Strength of } Fe^{2+} \text{ in the given solution} = \frac{55.85 \times V_1}{1000 \times 10} \times \frac{1000}{10} = x \text{ g/ litre}$$

Total Iron: 1000 ml of 1N $KMnO_4 \equiv 55.85$ g of Fe

$$\therefore \quad V_2 \text{ ml of } \frac{N}{10} KMnO_4 \equiv \frac{55.85}{1000} \times \frac{1}{10} \times V_2 \text{ g Fe}$$

$$\text{Total iron present in 10 ml. of the given solution} = \frac{55.85 \times V_2}{1000 \times 10} \cdot g$$

$\therefore$ Strength of total iron in the given solution

$$= \frac{55.85 \times V_2}{1000 \times 10} \times \frac{1000}{10} = y \text{ g/litre.}$$

Conclusion

The given solution contains

Total iron = y g/litre

Ferrous iron = x g/litre

Ferric iron = (y–x) g/litre.

Preparation of Reagents

1. *Potassium Permanganate (N/10):* Dissolve 3.1607 g of $KMnO_4$ (eq. wt. 31.607) in distilled water. Boil for 1 hour, cool, filter and makeup the volume to 1 litre with distilled water. Standardise it with oxalic acid to know the exact strength.
2. *Sulphuric Acid (5 N):* Add slowly 139 ml of conc. sulphuric acid (36 N, 98%) to distilled water. After cooling make up the volume of 1 litre with distilled water.
3. *Concentrated Solution of Stannous Chloride:* Dissolve 30 g of pure $SnCl_2 . 2H_2O$ in 100 ml of conc. HCl and dilute to 200 ml with distilled water.
4. *Mercuric Chloride (5%):* Dissolve 5 g of $HgCl_2$ in 100 ml of distilled water.

EXPERIMENT 3.6

Aim of the Experiment

Determination of sodium hydroxide and sodium carbonate in the given alkali mixture solution.

Apparatus Required

- Burette
- Pipette
- Conical flask
- Glazed tile.

Chemicals Required

- Given alkali mixture solution
- $\frac{N}{10}$ HCl solution
- Phenolphthalein indicator
- Methyl orange indicator.

Theory

Determination of NaOH and Na_2CO_3 when present together, double indicator method is generally employed. The principle involved in this method is that when sodium carbonate solution is titrated with hydrochloric acid, the neutralization occurs in two stages. The first one correspond to the hydrogen carbonate (HCO_3^-) stage, when the net reaction is,

$$CO_3^{2-} + H^+ \longrightarrow HCO_3^-$$

The equivalence point for the primary stage of ionization of carbonic acid is at pH 8.3, which may be detected by employing phenolphthalein indicator. The second stage of the titration corresponds to the displacement of all the carbonic acid, when the net reaction is,

$$CO_3^{2-} + 2H^+ \longrightarrow H_2CO_3$$
$$HCO_3^- + H^+ \longrightarrow H_2CO_3$$
$$H_2CO_3 \longrightarrow H_2O + CO_2$$

The equivalence point for the secondary stage of ionization is at pH range 3.1 to 4.4, which may be detected by employing methyl orange indicator.

Thus, when a mixture of Na_2CO_3 and NaOH is titrated with standard HCl, the phenolphthalein end point corresponds to the neutralisation of all the NaOH and partial neutralisation of all the Na_2CO_3 to the bicarbonate stage.

$$NaOH + HCl \longrightarrow NaCl + H_2O$$
$$Na_2CO_3 + HCl \longrightarrow NaCl + NaHCO_3$$

Now, if the titration is continued using methyl orange as indicator, the neutralisation of the bicarbonate to the carbonic acid occurs.

$$NaHCO_3 + HCl \longrightarrow NaCl + H_2CO_3$$
$$H_2CO_3 \longrightarrow H_2O + CO_2$$

So that methyl orange end point means the **total volume of the acid** run down from the beginning of the experiment.

Thus if [P] and [M] correspond to the phenolphthalein and methyl orange and pionts, then

$$[P] \equiv NaOH + \frac{1}{2} Na_2CO_3$$

and $$[M] \equiv NaOH + Na_2CO_3$$

Hence, $$[M] - [P] \equiv \frac{1}{2} Na_2CO_3$$

$$2\{[M] - [P]\} \equiv Na_2CO_3$$

and $$[M] - 2\{[M] - [P]\} \equiv NaOH.$$

Procedure

Pipette out 10 ml of given alkali mixture solution into a conical flask and add 2 to 3 drops of phenolphthalein indicator. The solution assumes pink colour. Titrate the solution with the $\frac{N}{10}$ HCl from the burette until the solution becomes colourless. Note down this titre value as phenolphthalein end point [P]. Then add 2 to 3 drops of methyl orange indicator to the same solution in the conical flask and continue the titration until a sharp colour change from yellow to red colour at the end point. Note down total titre value from the beginning of the experiment as methyl orange end point [M].

Tabulation

No. of observations	*Volume of alkali mixture solution (ml)*	*I.B.R. (ml)*	*Titration with Phenolphthalein*		*Titration with methyl orange*	
			F.B.R. (ml)	*Vol. of HCl used [P] (F.B.R.—I.B.R.)*	*F.B.R. (ml)*	*Vol. of HCl used [M] (F.B.R.–I.B.R.)*
1	10					
2	10					
3	10					

Calculation

NaOH Content: We know that equivalent weight of NaOH = 40

Hence, 1000 ml of 1N HCl ≡ 40 g of NaOH

$$1 \text{ ml of } 1 \text{ N HCl} \equiv \frac{40}{1000} \text{ g of NaOH}$$

$$\therefore \quad M - 2(M - P) \text{ ml of } \frac{N}{10} \text{ HCl} \equiv \frac{40}{1000} \times \frac{1}{10} \times M - 2(M - P) \text{ g of NaOH}$$

Strength of NaOH in the given solution

$$= \frac{40}{1000 \times 10} \times M - 2(M - P) \times \frac{1000}{10} = x \text{ g/litre}$$

Na_2CO_3 Content: We know that equivalent weight of Na_2CO_3 = 53

Hence 1000 ml of 1N HCl ≡ 53 g of Na_2CO_3

$$1 \text{ ml of 1N HCl} \equiv \frac{53}{1000} \text{ g of } Na_2CO_3$$

$$2\text{ (M–P) ml of } \frac{N}{10} \text{ HCl} \equiv \frac{53}{1000} \times \frac{1}{10} \times 2\text{ (M – P) g of } Na_2CO_3$$

Strength of Na_2CO_3 in the given solution

$$= \frac{53}{1000 \times 10} \times 2(\text{M} - \text{P}) \times \frac{1000}{10} = y \text{ g/litre.}$$

Conclusion

The given alkali mixture solution contains

x g/litre of NaOH

and y g/litre of Na_2CO_3.

Preparation of Reagents

1. *Hydrochloric Acid (N/10):* Dilute 8.4 ml of concentrated hydrochloric acid to 1 litre with distilled water and standardise with $\frac{N}{10}$ Na_2CO_3.
2. *Phenolphthalein Indicator:* Dissolve 1 g of phenolphthalein in 100 ml of distilled ethyl alcohol and add 100 ml of distilled water with constant stirring. Filter, if necessary.
3. *Methyl Orange Indicator:* Dissolve 50 mg of free acid form of methyl orange solid in distilled water and make the volume to 100 ml with distilled water. Filter the solution if necessary.

EXPERIMENT 3.7

Aim of the Experiment

Determination of percentage of available chlorine in a given sample of bleaching powder.

Apparatus Required

- Burette
- Pipette
- Volumetric flask
- Conical flask
- Funnel
- Pastle and mortar.

Chemicals Required

- Given sample of bleaching powder
- Potassium iodide
- Glacial acetic acid
- $\frac{N}{10}$ sodium thiosulphate solution (Hypo).

Theory

The major constituents of bleaching powder are calcium hypochlorite [$Ca(OCl)_2 . 4H_2O$], basic calcium chloride [$CaCl_2 . Ca(OH)_2 . H_2O$] and some free calcium hydroxide [$Ca(OH)_2$]. The active constituent of bleaching powder is hypochlorite ($CaOCl_2$) which is responsible for disinfection and bleaching action of bleaching powder. When $Ca(OCl)_2$ is treated with HCl, free chlorine is liberated according to the equation,

$$Ca(OCl)_2 + 4\,HCl \longrightarrow CaCl_2 + 2H_2O + 2Cl_2$$

$$OCl^- + Cl^- + 2H^+ \longrightarrow Cl_2 + H_2O$$

$$Cl_2 + H_2O \longrightarrow HCl \quad + \quad HOCl$$

Hypochlorous acid

↓

kills germs

$$Cl_2 + KI \longrightarrow KCl + I_2$$

$$I_2 + 2Na_2S_2O_3 \longrightarrow Na_2S_4O_6 + 2NaI$$

Sodium tetrathionate

The available chlorine refers to the chlorine liberated by the action of dilute acids and is expressed in percentage weight of bleaching powder. The liberated chlorine from bleaching powder sets free from KI an equivalent amount of iodine which is titrated against standard solution of hypo.

Procedure

Weigh out accurately 3 g of air dried bleaching powder in a clean, dried and preweighed weighing tube. Transfer the sample in a mortar and make a paste with little distilled water. Transfer the entire mass quantitatively to a 250 ml of volumetric flask and make upto the mark with distilled water. Pipette out 25 ml of the homogeneous solution from the volumetric flask to a conical flask. Add about 1 g of solid KI and about 20 ml of glacial acetic acid. Titrate the liberated iodine against the hypo solution till light yellow colour appears. Then add 10 drops of freshly prepared starch solution into it and continue the titration till the colour disappears. Repeat the titration to get at least two concordant readings. Note down the volume of hypo solution used (V_2 ml).

Tabulation

No. of observations	*Volume of $CaOCl_2$ solution (V_1 ml)*	*I.B.R. (ml)*	*F.B.R. (ml)*	*Volume of hypo solution used (Difference V_2 ml)*
1	25			
2	25			
3	25			

Calculation

Weight of the given bleaching powder sample = W g

N_1 = Normality of available chlorine = ?

V_1 = Volume of pipette out in conical flask = 25 ml

N_2 = Normality of the hypo solution = $\frac{N}{10}$

V_2 = Volume of hypo solution used = x ml (say)

$$N_1V_1 = N_2V_2$$

or,
$$N_1 \times 25 = \frac{N}{10} \times x$$

or,
$$N_1 = \frac{x}{250} N$$

Therefore amount of chlorine per litre of the solution = $\frac{x \times 35.5}{250}$ g

$$1000 \text{ ml } Ca(OCl)_2 \equiv \frac{x \times 35.5}{250}$$

$$1 \text{ ml } Ca(OCl)_2 \equiv \frac{x \times 35.5}{250 \times 1000}$$

$$250 \text{ ml } Ca(OCl)_2 \equiv \frac{x \times 35.5}{250 \times 1000} \times 250 \equiv \frac{x \times 35.5}{1000}$$

If W g of samle contains $\frac{x \times 35.5}{1000}$ g of Cl_2

100 g of sample contains $\frac{x \times 35.5}{1000} \times \frac{100}{W} = \frac{x \times 35.5}{W \times 10}$ g of Cl_2.

Conclusion

The percentage of available chlorine in the supplied sample of bleaching powder = $\frac{x \times 35.5}{W \times 10}$.

Preparation of Reagents

1. *Sodium Thiosulphate Solution (N/10):* Dissolve 24.82 g of $Na_2S_2O_3 \cdot 5H_2O$ (eq. wt. 248.21) in 1 litre of boiled out distilled water. Standardise it with $K_2Cr_2O_7$ to know the exact strength.
2. *Starch Indicator (1%):* Prepare a paste of 1 g of powdered starch with distilled water and add to this paste 100 ml of boiled water with constant stirring. Boil for about five mintues and then cool.

N.B. The approximate percentage of available chlorine in commercial bleaching powder is 36–38%.

EXPERIMENT 3.8

Aim of the Experiment

Estimation of nickel using Erio-T as indicator.

Apparatus Required

- Burette
- Pipette
- Conical flask
- Volumetric flask.

Chemicals Required

- Given nickel solution
- 0.1 M $ZnSO_4$ solution
- 0.1 M EDTA solution
- Buffer solution (pH 10)
- Eriochrome Black T indicator.

Theory

Nickel forms a very stable complex with EDTA, excess EDTA added may be back titrated with zinc ion solution using Eriochrome Black T indicator.

Procedure

Pipette out 20 ml of the neutral solution of nickel into a conical flask and add a little excess of 0.1 M EDTA solution (25–30 ml). Dilute the solution to about 200 ml with distilled water, add 4 ml of buffer solution (pH 10) followed by few drops of Eriochrome Black T indicator. Titrate the excess of EDTA with standard 0.1 M zinc sulphate solution, until the colour changes from blue to wine red. Repeat the experiment to get the concordant titre value (V_2 ml).

Tabulation

No. of observations	*Volume of EDTA solution (V_1 ml)*	*I.B.R. (ml)*	*F.B.R. (ml)*	*Volume of $ZnSO_4$ solution used (Difference V_2 ml)*
1				
2				
3				

Calculation

Every ml difference between the volume of 0.1 M EDTA solution added and the 0.1 M zinc ion solution used in the back titration corresponds to 5.871 mg of nickel.

Conclusion

The nickel content in the given solution is ______ ppm.

Preparation of Reagents

1. *EDTA Solution (0.1 M):* Weigh 37.225 g of disodium dihydrogen ethylene diamine tetra acetic acid (Mol. wt. 372.25) after getting dreid it in an oven at 75°C for 90 minutes and cooled. Dissolve it in distilled water in a volumetric flask and make up the volume to 1 litre with distilled water.
2. *Zinc Sulphate Solution (0.1 M):* Weigh 16.143 g of $ZnSO_4$ (Mol. wt. 161.43) and dissolve it in distilled water in a volumetric flask and make up the volume to 1 litre with distilled water.
3. *Buffer Solution (pH 10):* Dissolve 70 g of NH_4Cl and add 568 ml of conc. ammonia solution and dilute with distilled water to 1 L.
4. *Eriochrome Black T Indicator:* Dissolve 0.5 g of the dye stuff in 100 ml rectified spirit or methanol. The solution is stable for one month.

EXPERIMENT 3.9

Aim of the Experiment

Determination of copper by complexometric titration.

Apparatus Required

- Burette
- Pipette
- Conical flask
- Glazed tile.

Chemicals required

- Given copper solution
- 0.1 M EDTA solution
- Buffer solution (pH 10)
- Eriochrome Black T indicator.

Theory

$$Cu^{2+} + HD^{2-} \longrightarrow CuD^{-} + H^{+}$$

$$CuD^{-} + EDTA \longrightarrow Cu^{-}\ EDTA + HD^{2-}$$

Procedure

Pipette out 20 ml of the unknown Cu^{2+} solution into a 250 ml concial flask, dilute it by adding about 50 ml of distilled water, add 5 ml of buffer (pH 10) followed by 3–4 drops of Eriochrome Black T indicator solution. Titrate this solution against standard solution of EDTA from the burette until the colour changes from red to blue. Warm the solution to about 40°C for complete complexation to get better end point. Note down the final burette reading (V_2 ml). Repeat the titration to get at least two concordant readings.

Tabulation

No. of observations	*Volume of Cu^{2+} solution (V_1 ml)*	*I.B.R. (ml)*	*F.B.R. (ml)*	*Volume of EDTA solution used (Difference V_2 ml)*
1	20			
2	20			
3	20			

Calculation

$$\frac{M_1V_1}{n_1} = \frac{M_2V_2}{n_2}$$

M_1 = Molarity of Cu^{2+} solution = ?
V_2 = Volume of Cu^{2+} solution = 20 ml
n_1 = Mole equivalence of Cu^{2+}
M_2 = Molarity of EDTA solution
V_2 = Volume of EDTA solution = x ml (say)
n_2 = Mole equivalence of EDTA

$$M_1 = \frac{M_2V_2}{n_2} \times \frac{n_1}{V_1}$$

Amount of Cu^{2+} = M_1 × At. wt. of Cu
= M_1 × 63.54 g/L

Conclusion

Copper content in the given solution is M_1 × 63.54 g/L.

Preparation of Reagents

1. *EDTA Solution (0.1 M):* Weigh 37.225 g of disodium dihydrogen ethylene diamine tetra acetic acid (Mol. wt. 372.25) after getting dried it in an oven at 75°C for 90 minutes and cooled. Dissolve it in distilled water in a volumetric flask and make up the volume to 1 litre with distilled water.
2. *Buffer Solution (pH 10):* Dissolve 70 g of NH_4Cl and add 568 ml of conc. ammonia solution and dilute with distilled water to 1L.
3. *Eriochrome Black T Indicator:* Dissolve 0.5 g of the dye stuff in 100 ml rectified spirit or methanol. The solution is stable for one month.

EXPERIMENT 3.10

Aim of the Experiment

Determination of magnesium by complexometric titration.

Apparatus Required

- Burette
- Pipette
- Conical flask
- Glazed tile.

Chemicals Required

- Given magnesium solution ($MgSO_4$)
- 0.01 M EDTA solution
- Buffer solution (pH 10)
- Eriochrome Black T indicator.

Theory

EDTA in form of its disodium salt is used to estimate Mg^{2+} ions in the given $MgSO_4$ solution.

The indicator Eriochrome Black T forms an unstbale complex with Mg^{2+} solution taken in the flask at a pH 10. The complex is wine red in colour.

$$Mg^{2+} + EBT \longrightarrow \underset{\substack{\text{Unstable complex} \\ \text{(Wine red)}}}{(Mg) - EBT}$$

Once the free metal ions are complexed, the EDTA replaces Mg ions from the unstable complex and the indicator is set free. Since the indicator is blue at above pH, the end point is in appearance of blue colour.

$$\underset{\text{(Wine red)}}{[Mg - EBT]} + EDTA \longrightarrow \underset{\text{(Colourless)}}{[Mg - EDTA]} + \underset{\text{(Blue)}}{EBT}$$

Procedure

Pipette out 5 ml of given magnesium solution into the concial flask. To this add 5 ml buffer solution and dilute it with 10 ml of distilled water. Then add 2–3 dorps of EBT indicator. Titrate it against the standard EDTA solution taken in the burette till the colour changes from wine red to blue. Note down the titre value (V ml) repeat the titration to get concordant readings.

Tabulation

No. of observations	*Volume of $MgSO_4$ solution (ml)*	*I.B.R (ml)*	*F.B.R (ml)*	*Volume of EDTA solution used (Difference V ml)*
1	5			
2	5			
3	5			

Calculation

1 ml of 0.01 M EDTA = 0.243 mg of magnesium

V ml of 0.01 M EDTA = V × 0.243 mg of magnesium

∴ 5 ml of $MgSO_4$ solution contains V × 0.243 mg of magnesium

∴ 1000 ml of $MgSO_4$ solution contains

$$\frac{V \times 0.243}{5} \times 1000 \text{ mg/L of magnesium}$$

$$= \frac{V \times 0.243}{5} \times 1000 \text{ ppm of magnesium.}$$

Conclusion

The magnesium content in the given solution is $\frac{V \times 0.243}{5} \times 1000$ ppm.

Preparation of Reagents

1. *EDTA Solution (0.01 M):* Weigh 3.723 g of disodium dihydrogen ethylene diamine tetraacetic acid (Mol. wt. 372.25) after getting dried it in an oven at 75°C for 90 minutes and cooled. Dissolve it in distilled water in a volumetric flask and make up the volume to 1 litre with distilled water.
2. *Buffer Solution (pH 10):* Dissolve 70 g of NH_4Cl and add 568 ml of conc. ammonia solution and dilute with distilled water to 1 L.
3. *Eriochrome Black T Indicator:* Dissolve 0.5 g of the dye stuff in 100 ml rectified spirit or methanol. The solution is stable for one month.

3.3 GRAVIMETRIC ANALYSIS

Gravimetric analysis is a branch of quantitative analysis which deals with the estimation of compounds by measurement of weights. A known weight of the compound is treated with the solution of suitable reagent to get the precipitate of the known compound. It is filtered, washed, dried, ignited to constant weight. The weight of the final product is noted. From the weight of the original compound taken and of the final product formed, the amount of the original compound or mixture may be calculated. Several steps that include gravimetric analysis are preparation of the solution, precipitation, filtration, drying, ignition and weighing.

Thermo Gravimetric Analysis

In thermo gravimetric analysis a change in the weight of a substance is recorded as a function of temperature or time.

The basic instrumental requirement for thermo gravimetry is a precision balance with a furnace programmed for a linear rise of temperature with time. The results may be presented as a thermo gravimetric (TG) curve, in which weight change is recorded as a function of temperature or time.

EXPERIMENT 3.11

Aim of the Experiment

Estimation of silver as silver chloride.

Apparatus Required

- Beaker
- Pipette
- Watch glass

- Glass crucible (Porosity 3)
- Whatmann filter paper no. 40
- Electric even
- Desiccator.

Chemicals Required

- $AgNO_3$ solution (1%)
- 0.2 N HCl
- Dilute HNO_3 solution.

Theory

Silver chloride is precipitated in presence of dil. HNO_3 by adding slowly hot dil. HCl to $AgNO_3$ solution, with constant stirring.

$$AgNO_3 + HCl \xrightarrow{HNO_3} AgCl + HNO_3$$

$$Ag^+ + Cl^- \longrightarrow \underset{\text{(Precipitate)}}{AgCl}$$

The precipitate of AgCl is separated by filtration through a standard filter paper. It is washed with dilute HNO_3, dried and weighed. The amount and percentage of silver in the given sample may be calculated from the weight of the silver salt taken and the weight of dry AgCl is obtained by using the following relation,

$$\underset{108 \text{ parts}}{Ag} \equiv \underset{170 \text{ parts}}{AgNO_3} \equiv \underset{143.5 \text{ parts by weight}}{AgCl}$$

Procedure

Pipette out 20 ml of silver nitrate solution in a clean 250 ml beaker and add an equal volume of distilled water. Add 2.0 ml of conc. HNO_3 and cover the beaker with a thick brown paper to protect if from sunlight (sunlight can decompose AgCl to metallic silver).

Heat to about 70°C and add 0.2 N HCl to the soluton in small lots of 1–2 ml with stirring till precipitation is complete. Allow the precipitate to settle down and add few drops of 0.2 N HCl to the supernatant liquid. If there is no formation of precipitate, it shows complete precipitation. Heat the solution for 2–3 minutes to coagulate the precipitate. Cover the beaker with a watch glass and allow to stand for one hour so that the precipitate settles down.

Clean a sintered glass crucible of porosity 3 and dry it in an electric oven at 130–140°C. Cool it in a desiccator to room temperature and note its weight. Set the crucible for filtration and pour, the supernatant liquid into it by decantation. Wash the precipitate 2–3 times with dil. HNO_3. Now transfer the precipitate to the crucible with a spray of water from the wash bottle. Remove the precipitate with the help of a glass rod. Wash repeatedly with dil. HNO_3 till it is free from chloride ions by testing with $AgNO_3$ solution. Now keep the crucible containing the precipitate in an air oven at 130–150°C for half an hour. Cool in a desiccator to room temperature and note its weight.

Repeat the process of drying, cooling and weighing till constant weight is obtained.

Observation

Wt. of $AgNO_3$ dissolved per litre = 10 g

Wt. of empty crucible = W_1 g

Wt. of crucible + AgCl precipitate = W_2 g

Wt. of AgCl precipitate formed = $W_2 - W_1 = W$ g.

Calculation

Wt. of AgCl ppt. obtained from 20 ml solution = W g

$$\text{Wt. of AgCl ppt. obtained from 1000 ml} = \frac{1000 \times W}{20} = 50\ W\ g$$

$$\text{Wt. of } AgNO_3 \text{ present in 1000 ml of solution} = 50\ W \times \frac{170}{143.5} = X\ g\ (\text{say})$$

Now

$$\underset{143.5}{AgCl} \equiv \underset{108}{Ag}$$

$$\frac{\text{Mol. wt. of AgCl}}{\text{At. wt. of Ag}} = \frac{\text{Strength of AgCl}}{\text{Strength of Ag}}$$

or,

$$\frac{143.5}{108} = \frac{50\ W}{\text{Strength of Ag}}$$

$\therefore$

$$\text{Strength of Ag} = \frac{108 \times 50\ W}{143.5} = Y\ gm/L.$$

Preparation of Reagents

1. *Hydrochloric Acid (0.2 N):* Dilute 17 ml of concentrated hydrochloric acid (12 N) to 1 litre with distilled water and standardise with $\frac{N}{10}$ Na_2CO_3.
2. *Dilute Nitric Acid (5 N):* Dilute 309 ml of conc. nitric acid (16.2 N, 72%) to 1 L with distilled water.
3. *Silver Nitrate Solution (1%):* Dissolve 1 g of $AgNO_3$ in 100 ml of distilled water.

EXPERIMENT 3.12

Aim of the Experiment

Estimation of barium in barium sulphate ($BaSO_4$).

Apparatus Required

- Beaker
- Watch glass
- Funnel

- Sintered glass crucible (Porosity 4)
- Whatmann filter paper no. 42
- Muffle furnace
- Desiccator.

Chemicals Required

- Given $BaCl_2$ solution
- 1 N H_2SO_4
- Conc. HCl
- $AgNO_3$ solution (N/50).

Thoery

Barium is precipitated as barium sulphate by adding slowly hot dil. H_2SO_4 to hot solution of barium salt acidified slightly with HCl.

$$BaCl_2 + H_2SO_4 \longrightarrow BaSO_4 \downarrow + 2HCl$$

The precipitate of $BaSO_4$ is separated by filtration either through whatmann filter paper 42 or a sintered glass filtration crucible of porostiy 4. It is then washed, dried, ignited and weighed. Knowing the weight of the barium salt dissolved and precipitate of $BaSO_4$ formed, the amount of barium salt may be estimated as follows.

$$\underset{233 \text{ parts}}{BaSO_4} \equiv \underset{137 \text{ parts}}{Ba^{2+}} \equiv \underset{208 \text{ parts}}{BaCl_2} \equiv \underset{208 + 18x \text{ parts by wt.}}{BaCl_2 \,.\, x\, H_2O}$$

The precipitation is carried out in the presence of dil. HCl to avoid the precipitation of carbonates and phosphates present in the solution as impurities and also to make precipitate granular.

Procedure

Pipette out 20 ml of the given solution of barium salt in a 250 ml beaker provided with a stirring rod and covered with a watch glass. Dilute the solution to about 100 ml with distilled water.

Add about 1 ml of conc. HCl to the solution drop by drop and heat the solution to boiling keeping the beaker covered with a watch glass.

Remove the beaker from the burner. Rinse the watch glass with distilled water into the solution. Now, add 1N H_2SO_4 drop by drop to the solution with constant stirring till the precipitation is complete.

Allow the precipitate to settle down until the supernatant liquid is clear. Now add a few drops of dil. H_2SO_4 in the supernatant liquid without disturbing the precipitate and see for complete precipitation. If precipitation is not completed add more precipitant (1N H_2SO_4) for complete precipitation.

Now cover the beaker with a watch glass and digest it over a water bath for about half an hour till the precipitate settles down. Now set the Whatmann filter paper no. 42 on a funnel placed on a tripod stand. Add clear supernatant liquid to the filter paper. Wash the precipitate 2–3 times with hot water and pour the washings into the filter paper. Transfer the precipitate from the beaker into the filter paper. Remove the particles sticking to the sides by a glass rod. Wash the precipitate with hot water till it is free from chloride ions by testing with $AgNO_3$ solution.

When the filter paper is completely drained, fold the filter paper to form a packet of the precipitate. Remove from the funnel and place in a tarred and weighed cruicible placed in a clay triangle on tripod stand. Keep it half covered with a lid in slanting position. Heat with a bunsen burner on a low flame till it is dried. Raise the temperature to char the filter paper and volatile matter is expelled. Continue heating till all carbon is converted into ash.

Cool the crucible and treat the ash with one drop each of conc. HCl and the conc. H_2SO_4. Heat first slowly and then strongly with care, till there is no evolution of white fumes. Cool the crucible and place it in a muffle furnace at 600–800°C for half an hour for ignition till complete ashing. Cool the crucible first in air and then in a desiccator to room temperature and weigh accurately.

Repeat the process of heating, cooling and weighing till a constant weight is obtained.

Observations

Volume of solution taken = 20 ml

Wt. of empty crucible = W_1 g

Wt. of crucible + ppt. of $BaSO_4$ = W_2 g

Wt. of $BaSO_4$ precipitate = $W_2 - W_1 = W$ g

Weight of $BaSO_4$ ppt. obtained from 20 ml solution = W g

Weight of $BaSO_4$ ppt. obtained from 1000 ml of solution = $W \times \frac{1000}{20} = 50\ W$

$\therefore$ Wt. of $BaSO_4 = \frac{137.36}{233.42} \times W$ g of Ba^{2+} ions

W is the wt. of $BaSO_4$ obtained from 20 ml of the solution.

Hence, the amount of Ba^{2+} ions per litre of the given solution,

$$= \frac{137.36 \times 50 \times W}{233.42}\ g = X\ g$$

Conclusion

The amount of Ba^{2+} ions in the given solution is X g/L.

Preparation of Reagents

1. *Sulphuric acid (1N):* Add 27.80 ml of sulphuric acid (36 N, 98%) to distilled water. After cooling make up the volume to 1L with distilled water.
2. *Silver Nitrate Solution* $\left(\frac{N}{50}\right)$: Dissolve 3.3978 g of silver nitrate solid (eq. wt. 169.89) in 1L of distilled water.

VIVA-VOCE

1. Define volumetric analysis.
2. What is called end point of a titration?
3. What is difference between end point and equivalence point?
4. What is standard solution?
5. What do you mean by primary and secondary standard?
6. What do you mean by standardisation?
7. What is an external indicator?
8. Which indicator is used in $KMnO_4$ titration?
9. Why $KMnO_4$ is acidified with only sulphuric acid and not hydrochloric acid or nitric acid?

   Ans. Nitric acid is a strong oxidising agent and hydrochloric acid a strong reducing agent, hence both react with $KMnO_4$. So, $KMnO_4$ is acidified with only H_2SO_4 and not HNO_3 or HCl.*
10. Why the oxalic acid solution is heated before titration with $KMnO_4$ solution?

    Ans. The reaction of oxalic acid with $KMnO_4$ is very slow. Therefore the oxalic acid soltuion is heated to 60–70°C initially. Once the reaction has started, its rate automatically increases due to formation of Mn^{2+} ions which catalyse the reaction (auto catalysis). The solution should not be allowed to boil because oxalic acid decomposes on boiling.
11. What is Mohr's salt?
12. What is hypo?
13. What is the trade name of EDTA?
14. Why do we use the disodium salt of EDTA?

    Ans. As disodium salt is soluble in water.
15. What is the difference between iodimetry and iodometry?
16. What is the formula of bleaching powder?
17. What is available chlorine in bleaching powder?
18. Explain the mechanism of sterilization of water by bleaching powder.
19. What is the other name of bleaching powder?

    Ans. The other name of bleaching powder is chloride of lime or bleach.
20. What is a buffer solution?
21. What is the pH range of phenolphthalein and methyl orange?
22. What is gravimetric analysis?

Chapter 4

WATER AND WASTE WATER ANALYSIS

The planet earth is abudantly rich in water, which constitute 75% of its surface. Most of the water which is available is located in oceans (97.02%). The high salinity makes it unusable as a source of water for municipal, agricultural and industrial purposes. Hence we are left with only 2.98% of water, out of whcih considerable amount of water is present in the form of polar ice caps and many remain locked as such for thousands of years. The conclusion is that only 1% of world's water resources is found as fresh water for human consumption and other uses. Water is used for industrial purposes and for municipal supply. It is necessary to ensure the right quality and quantity of water for these purposes. Treatment of water obtained from different sources is therefore essential so that it may be used safely for industrial or municipal purposes.

4.1 ANALYSIS OF WATER

Analysis of water includes the determination of different constituents in water to ascertain the suitability of water for a specific use. The following characteristics of water are usually determined.

- Hardness
- Alkalinity
- Dissolved oxygen
- Biological oxygen demand
- Chemical oxygen demand
- Free chlorine
- Chloride.

EXPERIMENT 4.1

Aim of the Experiment

Determination of total hardness of water by EDTA method.

Apparatus Required

- Burette
- Pipette
- Beaker
- Conical flask
- Volumetric flask
- Funnel.

Chemicals Required

- Standard hard water
- 0.01 M EDTA solution
- Buffer solution
- Eriochrome Black T indicator (EBT)

Theory

Hardness of water is the property by virtue of which it prevents the lathering of soap. Hardness occurs due to the presence of certain salts of Ca and Mg ions in water. These react with the sodium salts of long chain fatty acids present in the soap and are precipitated out in the form of calcium and magnesium stearate, palmitate or oleates. No lather is formed until all these ions are completely removed and thus a large amount of soap is wasted.

$$\underset{\text{Sodium stearate}}{2C_{17}H_{35}COONa} + CaCl_2 \longrightarrow \underset{\text{Calcium stearate}}{(C_{17}H_{35}COO)_2\,Ca} \downarrow + 2NaCl$$

Hardness is of two types:

1. *Temporary or Carbonate Hardness:* It is caused by the presence of carbonate and bicarbonate of calcium and magnesium and is removed on boiling. This is also called alkaline hardness.

$$Ca(HCO_3)_2 \xrightarrow{\Delta} CaCO_3 + H_2O + CO_2$$

2. *Permanent or Non-carbonate Hardness:* It is caused by the presence of chlorides and sulphates of calcium and magnesium which can not be removed simply by boiling. This is also called non-alkaline hardness.

The hardness of water may be determined by complexometric titration. In aqueous solution EDTA ionises to give two Na^+ ions and a strong chelating agent.

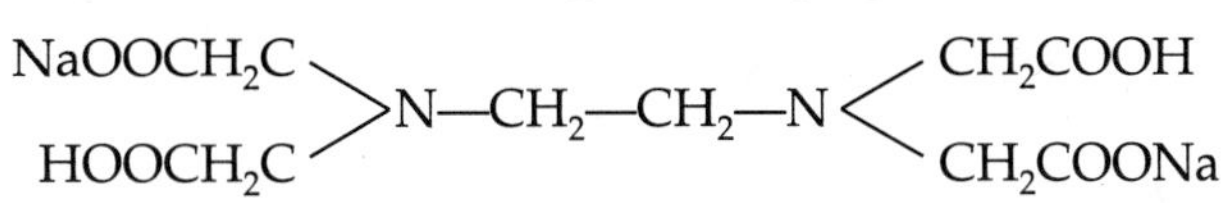

Disodium salt of ethylene diamine tetraacetic acid (EDTA)

The titration is carried out in presence of indicator Eriochrome Black T which is a complex organic compound.

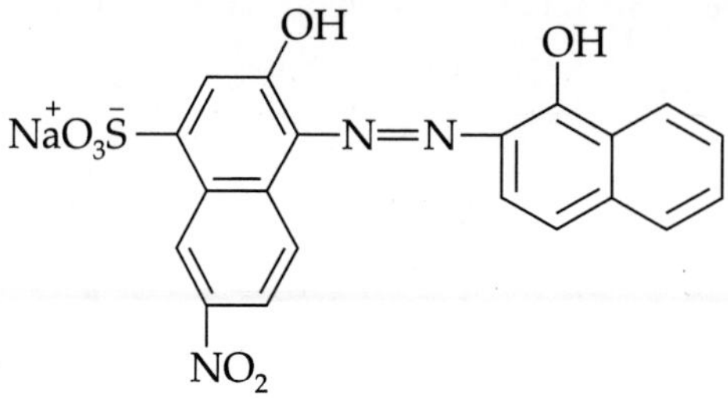

Eriochrome Black T

Sodium-1(1 hydroxy 2-naphthyl azo)-6-nitro-2-naphthol-4-sulphonate

The indicator when added in small amount to hard water buffered to a pH of about 10 combines with a few of Ca^{2+}, Mg^{2+} ions to form a weak complex of wine red colour which changes to blue when an excess drop of EDTA is added.

$$\begin{bmatrix} Ca^{2+} \\ Mg^{2+} \end{bmatrix} + EBT \longrightarrow \underset{\substack{\text{Unstable complex} \\ \text{(wine red)}}}{\begin{bmatrix} Ca^{2+} \\ Mg^{2+} \end{bmatrix} + EBT} \xrightarrow{EDTA} \underset{\text{Complex}}{\begin{bmatrix} Ca^{2+} \\ Mg^{2+} \end{bmatrix} + EDTA} + \underset{\text{(Blue)}}{EBT}$$

Thus, the change of wine red colour to distinct blue marks the end point of titration. Hardness of water is expressed as calcium carbonate equivalent.

Procedure

1. *Standardisation of EDTA Solution:* Take 25 ml of standard hard water in a conical flask. Add 10 ml buffer solution to it followed by 2–3 drops of Eriochrome Black T indicator. The colour of the solution becomes wine red. Titrate the solution against standard EDTA solution from the burette until the colour changes from wine red to blue at the end point. Take at least two concordant readings and note the titre value (V_1 ml).

2. *Determination of Total Hardness:* Take 25 ml of the supplied hard water in a conical flask. Add 10 ml buffer solution in to it followed by 2–3 drops of Eriochrome Black T indicator. Titrate the solution slowly against EDTA solution from the burette till the end point is reached when the colour changes from wine red to blue. Take two concordant readings and note the titre value (V_2 ml).

3. *Determination of Permanent Hardness:* Take 25 ml of supplied hard water in a 50 ml beaker and boil gently for about 10 minutes, cool and filter into a 25 ml of volumetric flask and make the volume upto mark with distilled water. Take this solution in a conical flask. Add 10 ml buffer solution into it followed by 2–3 drops of EBT indicator. Titrate the solution slowly against EDTA solution from the burette till the colour changes from wine red to blue. Take two concordant readings and note the titre value (V_3 ml).

Tabulation

Standardisation of EDTA solution

No. of observations	*Volume of standard hard water (ml)*	*I.B.R. (ml)*	*F.B.R. (ml)*	*Volume of EDTA solution used (Difference V_1 ml)*
1	25			
2	25			
3	25			

Total Hardness

No. of observations	*Volume of supplied hard water (ml)*	*I.B.R. (ml)*	*F.B.R. (ml)*	*Volume of EDTA solution used (Difference V_2 ml)*
1	25			
2	25			
3	25			

Permanent Hardness

No. of observations	*Volume of supplied hardwater after treatment for removal of temporary hardness (ml)*	*I.B.R. (ml)*	*F.B.R. (ml)*	*Volume of EDTA solution used (Difference V_3 ml)*
1	25			
2	25			
3	25			

Calculation

Standardisation of EDTA Solution: 1 ml standard hard water = 1 mg $CaCO_3$

$$V_1 \text{ ml EDTA} = 25 \text{ ml standard hard water}$$

$$= 25 \text{ mg } CaCO_3$$

$$1 \text{ ml EDTA} = \frac{25}{V_1} \text{ mg } CaCO_3.$$

Total Hardness: 25 ml hard water = V_2 ml EDTA

$$= V_2 \times \frac{25}{V_1} \text{ mg } CaCO_3$$

$$1 \text{ ml hard water} = V_2 \times \frac{25}{V_1 \times 25} \text{ mg } CaCO_3$$

$$1000 \text{ ml hard water} = V_2 \times \frac{25}{V_1} \times \frac{1000}{25} \text{ CaCO}_3$$

$$= \frac{V_2}{V_1} \times 1000 \text{ CaCO}_3$$

$$\text{Total hardness} = \frac{V_2}{V_1} \times 1000 \text{ ppm} = \text{X ppm (say).}$$

Permanent Hardness: 25 ml hard water = V_3 ml EDTA

$$= V_3 \times \frac{25}{V_1} \text{ mg CaCO}_3$$

$$1000 \text{ ml hard water} = V_3 \times \frac{25}{V_1} \times \frac{1000}{25} \text{ mg CaCO}_3$$

$$= \frac{V_3}{V_1} \times 1000 \text{ mg CaCO}_3$$

$$\text{Permanent hardness} = \frac{V_3}{V_1} \times 1000 \text{ ppm}$$

Temporary hardness = Total hardness – Permanent hardness

$$= \left(\frac{V_2}{V_1} - \frac{V_3}{V_1}\right) \times 1000 \text{ ppm.}$$

Conclusion

Total hardness of supplied hard water = X ppm.

Preparation of Reagents

1. *Standard Hard Water:* Dissolve 1.0 g of pure dry $CaCO_3$ in small amount of dilute hydrochloric acid and then evaporate the solution to dryness on a water bath. Dissolve the residue in distilled water to make 1 L solution (1 ml ≡ 1 mg $CaCO_3$).
2. *EDTA Solution (0.01 M):* Weigh 3.722 g of disodium dihydrogen ethylene diamine tetra acetic acid (Mol. wt. 372.25) after getting dried it in an oven at 75°C for 90 minutes and cooled. Dissolve it in distilled water in a volumetric flask and make up the volume to 1L with distilled water.
3. *Buffer Solution (pH 10):* Dissolve 70 g of NH_4Cl and add 568 ml of conc. ammonia solution and dilute with distilled water to 1L.
4. *Eriochrome Black T Indicator:* Dissolve 0.5 g of the dye stuff in 100 ml rectified spirit or methanol. The solution is stable for one month.

EXPERIMENT 4.2

Aim of the Experiment

Determination of total hardness of water by soap titration method (Clarke's method).

Apparatus Required

- Burette
- Pipette
- Conical flask
- Beaker
- Volumetric flask.

Chemicals Required

- Soap solution
- Standard hard water.

Theory

The soap titration method gives approximate results in estimating hardness. Hard water when treated with soap does not form lather. The calcium and magnesium ions present in water react with soap (sodium or potassium salts of higher fatty acids) to form insoluble precipitates of calcium or magnesium salt.

$$\underset{\text{Soap}}{2C_{17}H_{35}COONa} + Ca(HCO_3)_2 \longrightarrow \underset{\text{(Insoluble)}}{(C_{17}H_{35}COO)_2\,Ca} \downarrow + 2NaHCO_3$$

$$2C_{17}H_{35}COONa + CaSO_4 \longrightarrow (C_{17}H_{35}COO)_2\,Ca \downarrow + Na_2SO_4$$

$$2C_{17}H_{35}COONa + MgSO_4 \longrightarrow (C_{17}H_{35}COO)_2\,Mg \downarrow + Na_2SO_4$$

No lather is formed until all these ions are completely removed and thus a large amount of soap is consumed. Thus, the amount of soap initially consumed before the formation of permanent lather is a measure of total hardness or the amount of calcium and magnesium salts present in the hard water. A permanent lather is the lather which lasts for at least one minute after the mixture has been shaken. Hence the total hardness of a water sample can be determined by titrating the water sample against a soap solution of known strength. The end point is marked by the appearance of a stable lather at least for two minutes.

To find out the permanent hardness of the sample, the water sample is boiled for half an hour to remove the temporary hardness and the filtered water was treated with standard soap solution. The titre value corresponds to the permanent hardness of water.

Lather Factor: Lather factor is defined as the amount of standard soap solution required to produce a permanent lather in 50 ml of distilled water. It should be noted here that when soap is treated with distilled water (water of zero hardness), lather is not formed until some critical lower value of surface tension is reached.

The value of lather factor varies from 0.5 to 1.4 ml according to the kind of soap used. This value is subtracted from all the titre values obtained for hard water.

Procedure

1. *Determination of Lather Factor:* Pipette out 50 ml of distilled water (zero hardness) in a conical flask. Titrate it with soap solution taken in the burette till the end point is reached when stable lather persists for about 2 minutes. Take two concordant readings and note down the volume of soap solution used (V_1 ml).
2. *Standardisation of Soap Solution:* Pipette out 50 ml of standard hard water in the conical flask. Titrate it with soap solution taken in the burette till the end point is reached when a stable lather is formed over the entire surface of the water which remains for two minutes. Take atleast two concordant readings and note down the volume of soap solution used (V_2 ml).
3. *Determination of Total Hardness:* Pipette out 50 ml of the supplied hard water in the conical flask. Titrate it with soap solution taken in the burette till the end point is reached when a stable lather is formed over the entire surface of the water which remains for two minutes. Take atleast two concordant readings and note down the volume of soap solution used (V_3 ml).

Tabulation

Lather Factor

No. of observations	*Volume of distilled water (ml)*	*I.B.R. (ml)*	*F.B.R. (ml)*	*Volume of soap solution used (Difference V_1 ml)*
1	50			
2	50			
3	50			

Standardisation of Soap Solution

No. of observations	*Volume of standard hard water (ml)*	*I.B.R. (ml)*	*F.B.R. (ml)*	*Volume of soap solution used (Difference V_2 ml)*
1	50			
2	50			
3	50			

Total Hardness

No. of observations	*Volume of supplied hard water (ml)*	*I.B.R. (ml)*	*F.B.R. (ml)*	*Volume of soap solution used (Difference V_3 ml)*
1	50			
2	50			
3	50			

Calculation

Strength of Soap Solution: Soap solution consumed by 50 ml of standard hard water

$$= (V_2 - V_1) \text{ ml.}$$

Soap solution consumed by 50 mg of $CaCO_3$ = $(V_2 - V_1)$ ml

Since, 1 ml of standard hard water ≡ 1 mg of $CaCO_3$

$(V_2 - V_1)$ ml of soap solution ≡ 50 mg of $CaCO_3$

$$1 \text{ ml of soap solution} \equiv \frac{50}{V_2 - V_1} \text{ mg of } CaCO_3.$$

Total Hardness: 50 ml of sample hard water ≡ $(V_3 - V_1)$ ml of soap solution

$$\therefore \quad 50 \text{ ml of sample hard water} \equiv V_3 - V_1 \times \frac{50}{V_2 - V_1} \text{ mg of } CaCO_3$$

1000 ml of sample hard water contains

$$= V_3 - V_1 \times \frac{50}{V_2 - V_1} \times \frac{1000}{50} \text{ mg of } CaCO_3$$

$$= \frac{V_3 - V_1}{V_2 - V_1} \times 1000 \text{ mg of } CaCO_3$$

$$\therefore \qquad \text{Total hardness} = \frac{V_3 - V_1}{V_2 - V_1} \times 1000 \text{ ppm.}$$

Conclusion

The total hardness of the supplied hard water

$$= \frac{V_3 - V_1}{V_2 - V_1} \times 1000 \text{ ppm.}$$

Preparation of Reagents

1. *Soap Solution:* Shake vigorously 100 g of pure powdered castile soap in 1 L of 80% alcohol. Allow this solution to stand at least overnight, decant off the clear solution for standardisation.
2. *Standard Hard Water:* Dissolve 1.0 g of pure dry $CaCO_3$ in small amount of dilute hydrochloric acid and then evaporate the solution to dryness on a water bath. Dissolve the residue in distilled water to make 1 L solution (1 ml ≡ 1 mg $CaCO_3$).

EXPERIMENT 4.3

Aim of the Experiment

Determination of alkalinity of a given water sample.

Apparatus Required

- Burette
- Pipette
- Conical flask
- Glazed tile.

Chemicals Required

- $\frac{N}{10}$ HCl solution
- Phenolphthalein indicator
- Methyl orange indicator.

Theory

Alkalinity of water, which is a measure of the ability of water to neutralise the acids is due to the presence of bicarbonates, carbonates and hydroxides of calcium and magnesium. Determination of alkalinity due to different ions is based on the titration of the water sample against a standard acid making selective use of indicators. The indicators used are phenolphthalein and methyl orange. The reactions taking place are as follows:

$$\left.\begin{array}{l}\left.\begin{array}{l}1.\ OH^- + H^+ \longrightarrow H_2O \\ 2.\ CO_3^{2-} + H^+ \longrightarrow HCO_3^-\end{array}\right] P \\ 3.\ HCO_3^- + H^+ \longrightarrow H_2CO_3\end{array}\right] M$$

A known volume of the sample is titrated against a standard acid using phenolphthalein as indicator to phenolphthalein end point [P] and continuing the titration using methyl orange as indicator to methyl orange end point [M]. The volume of acid run down upto phenolphthalein end point [P] corresponds to the completion of equations (1) and (2) given above, while the volume of acid rundown after [P] corresponds to the completion of equation (3). The total amount of acid used from the beginning of the experiment, *i.e.* [M] corresponds to completion of the reaction (1) to (3).

The results may be summarised in the following table from which the alkalinity due to different ions can be calculated.

Alkalinity Due to Different Ions

Titration result	OH^-	CO_3^{2-}	HCO_3^-
[P] = 0	NIL	NIL	[M]
[P] = [M]	[P] or [M]	NIL	NIL
[P] = $\frac{1}{2}$ [M]	NIL	2[P] or [M]	NIL
[P] > $\frac{1}{2}$ [M]	2[P] – [M]	2{[M] – [P]}	NIL
[P] < $\frac{1}{2}$ [M]	NIL	2[P]	[M] – 2[P]

Alkalinity is generally expressed as parts per million in terms of $CaCO_3$.

Procedure

Take 50 ml of water sample in a 250 ml conical flask. Add 2–3 drops of phenolphthalein indicator. The colour of the solution will become pink. Now titrate this solution against 0.1 N HCl taken in a burette till colour of the solution disappears. It shows all the carbonates have been converted into bicarbonates. Note the titre value of the phenolphthalein end point [P]. Add 2–3 drops of methyl orange indicator to the same solution and continue the titration until a sharp colour change from yellow to rose red takes place. Note the total titre value from the beginning of the experiment as methyl orange end point [M]. Repeat the steps to get ioncordant readings.

Tabulation

No. of observations	*Volume of water sample (ml)*	*I.B.R. (ml)*	*Titration with phenolphthalein*		*Titration with methyl orange*	
			F.B.R (ml)	*Vol. of HCl used [P] (F.B.R.—I.B.R.)*	*F.B.R. (ml)*	*Vol. of HCl used [M] (F.B.R.–I.B.R.)*
1	50					
2	50					
3	50					

Calculation

Volume of water sample taken = 50 ml

Strength of HCL = 0.1 N

1 ml of 0.1 N HCl = 0.003 g CO_3^{2-}

= 0.0061 g HCO_3^-

= 0.005 g $CaCO_3$

= 0.0081 g $Ca(HCO_3)_2$

1. Total alkalinity as sodium carbonate in ppm

$$= [M] \times \text{Normality of acid} \times 0.053 \times \frac{10^6}{\text{Volume of water sample taken in ml}}$$

To calculate as $CaCO_3$ use the factor 0.05005 instead of 0.053.

2. Hydroxide alkalinity as OH^- radical in ppm

$$= (2[P] - [M]) \times \text{Normality of acid} \times 0.01701 \times \frac{10^6}{\text{Vol. of water sample taken in ml}}$$

To calculate as NaOH use the factor 0.040 instead of 0.01701.

3. Bicarbonate alkalinity as HCO_3^- radical in ppm

$$= ([M] - 2[P]) \times \text{Normality of acid} \times 0.06102 \times \frac{10^6}{\text{Vol. of water sample taken in ml}}$$

To calculate as Na_2CO_3, use the factor 0.084 instead of 0.06102.

To calculate as $CaCO_3$, use the factor 0.05005 instead of 0.06102.

4. Carbonate alkalinity as CO_3^{2-} in presence of OH^- in ppm

$$= 2([M] - [P]) \times \text{Normality of acid} \times 0.030 \times \frac{10^6}{\text{Vol. of water sample taken in ml}}$$

To calculate as Na_2CO_3 use the factor 0.053, instead of 0.030.

To calculate as $CaCO_3$ use the factor 0.05005 instead of 0.030.

5. Carbonate alkalinity as CO_3^{2-} in absence of hydroxide in ppm

$$= 2[P] \times \text{Normality of acid} \times 0.030 \times \frac{10^6}{\text{Vol. of water sample taken in ml}}.$$

Conclusion

The supplied water sample contains

Total alkalinity = ... ppm

Alkalinity with respect to CO_3^{2-} = ... ppm

Alkalinity with respect to HCO_3^- = ... ppm

Alkalinity with respect to OH^- = ... ppm.

Preparation of Reagents

1. *Hydrochloric Acid (N/10):* Dilute 8.4 ml of concentrated hydrochloric acid to 1 L with distilled water and standardise with $\frac{N}{10}$ Na_2CO_3.
2. *Phenolphthalein Indicator:* Dissolve 1 g of phenolphthalein in 100 ml. of distilled ethyl alcohol and add. 100 ml of distilled water with constant stirring, if necessary.
3. *Methyl Orange Indicator:* Dissolve 50 mg of free acid form of methyl orange solid in distilled water and make the volume to 100 ml with distilled water. Filter the solution if necessary.

EXPERIMENT 4.4

Aim of Experiment

Determination of dissolved oxygen in a given water sample (Winkler's method).

Apparatus Required

- Glass stoppered bottle
- Burette
- Pipette

- Conical flask
- Measuring cylinder
- Glazed tile.

Chemicals Required

- $\frac{N}{40}$ $Na_2S_2O_3$ solution (Hypo)
- $MnSO_4$ (48%)
- Alkaline potassium iodide
- Conc. H_2SO_4
- Starch.

Theory

Deficiency of oxygen in water is a kind of water pollution. Dissolved Oxygen (DO) is needed for living organisms to maintain their biological processes. In the absence of sufficient amount of dissolved oxygen in water, the anaerobic degradation of the pollutants make the water foul smelling. DO in boiler water is responsible for boiler corrosion. DO test helps to assess the raw water quality for different kinds of its utility.

Determination of dissolved oxygen in a sample of water is based upon the oxidation of KI by dissolved oxygen present in water sample. The liberated iodine is titrated against a standard hypo solution using starch as indicator. Dissolved molecular oxygen in water is not capable to react with KI, therefore an oxygen carrier such as manganese hydroxide is used. $Mn(OH)_2$ is produced by the action of KOH on $MnSO_4$. $Mn(OH)_2$ so obtained reacts with dissolved molecular oxygen to form a brown precipitate of basic manganic oxide $MnO(OH)_2$ which reacts with concentrated H_2SO_4 to liberate nascent oxygen. Nascent oxygen results oxidation of KI to I_2. This liberated I_2 is then titrated against standard $Na_2S_2O_3$ solution using starch as indicator. The reactions involved are:

$$MnSO_4 + 2KOH \longrightarrow Mn(OH)_2 + K_2SO_4$$

$$2Mn(OH)_2 + O_2 \longrightarrow \underset{\text{Basic manganic oxide}}{2MnO(OH)_2}$$

$$MnO(OH)_2 + H_2SO_4 \longrightarrow MnSO_4 + 2H_2O + \underset{\text{Nascent oxygen}}{[O]}$$

$$2KI + H_2SO_4 + [O] \longrightarrow K_2SO_4 + H_2O + I_2$$

$$2Na_2S_2O_3 + I_2 \longrightarrow Na_2S_4O_6 + 2NaI$$

Starch forms a deep blue colour, starch iodide complex when the solution turns blue. At the end point the blue colour disappears and the solution becomes colourless.

The nitrites present in water interfere with the titration as these can also liberate I_2 from KI.

$$2HNO_2 + H_2SO_4 + 2KI \longrightarrow 2NO + K_2SO_4 + 2H_2O + I_2$$

Sodium azide is used which decomposes nitrites into nitrogen.

$$2NaN_3 + H_2SO_4 \longrightarrow \underset{\text{Hydrazoic acid}}{2HN_3} + Na_2SO_4$$

$$HNO_2 + HN_3 \longrightarrow N_2O + N_2 + H_2O.$$

Procedure

Collect 300 ml of water sample in a glass stoppered bottle, avoiding contact with air. Add 2 ml of 48% $MnSO_4$ followed by 3 ml of alkaline potassium iodide reagent by means of pipette. Stopper the bottle and shake vigorously. Allow the brown precipitates to settle down. Now add 1 ml of conc. H_2SO_4 and mix until the precipitate is completely dissolved. Measure 100 ml of this solution by a measuring cylinder into a conical flask and titrate slowly against N/40 hypo solution, until the solution becomes pale yellow. Add about 2 ml of starch solution (freshly prepared) and continue the titration till the blue colour disappears. Note the titre value (V ml). Repeat the titration to get at least two concordant readings.

Tabulation

No. of observations	*Volume of solution taken in the conical flask (ml)*	*I.B.R. (ml)*	*F.B.R. (ml)*	*Volume of hypo solution used (Difference V ml)*
1	100			
2	100			
3	100			

Calculation

$$1000 \text{ ml of } 1 \text{ N } Na_2S_2O_3 \equiv 8000 \text{ mg of } O_2$$

$$1 \text{ ml of } 1 \text{ N } Na_2S_2O_3 \equiv 8 \text{ mg of } O_2$$

$$V \text{ ml of } \frac{N}{40} Na_2S_2O_3 \equiv 8 \times V \times \frac{1}{40} \text{ mg of } O_2$$

$$100 \text{ ml of the solution contain } 8 \times V \times \frac{1}{40} \text{ mg of } O_2$$

$$1000 \text{ ml of the solution contain } 8 \times V \times \frac{1}{40} \times \frac{1000}{100}$$

$$= X \text{ mg of } O_2 \text{ (say)}.$$

Conclusion

The amount of dissolved oxygen in the given water sample = X ppm.

Preparation of Reagents

1. *Sodium Thiosulphate Solution (N/40):* Dissolve 6.20 g of $Na_2S_2O_3$ solution in distilled water and make up this volume to 1 L. Standardise it with $\frac{N}{40}$ $K_2Cr_2O_7$ solution to know the exact strength.

2. *Manganese Sulphate Solution (48%):* Dissolve 480 g of $MnSO_4.4H_2O$ in distilled water and dilute to 1L.
3. *Alkaline Potassium Iodide Solution:* Dissolve 700 g of potassium hydroxide and 150 g of potassium iodide in distilled water. Cool and dilute to 1 L.
4. *Starch Indicator (1%):* Prepare a paste of 1 g of powdered starch with distilled water and add to this paste 100 ml. of boiled water with constant stirring. Boil for about five minutes and then cool.

EXPERIMENT 4.5

Aim of the Experiment

Determination of Biochemical Oxygen Demand (BOD) of a given waste water sample.

Apparatus Required

- BOD bottle
- Burette
- Pipette
- Volumetric flask
- Conical flask
- Incubator.

Chemicals Required

- Magnesium sulphate solution (22.5 g/L)
- Calcium chloride solution (27.5 g/L)
- Ferric chloride solution (0.25 g/L)
- Phosphate buffer solution.

Theory

BOD is a measure of the amount of oxygen required for the biological oxidation of the organic matter under aerobic conditions at 20°C and for a period of 5 days and is directly related to the extent of pollution of waste water, sewage etc. This is a measure of water pollution.

Sufficient oxygen must be present or supplied externally to maintain aerobic conditions so that organic matter decomposes into acceptable products. In anaerobic (in absence of oxygen) conditions, decomposition results highly objectionable end products including H_2S, NH_3 and CH_4.

The BOD test is based upon determination of Dissolved Oxygen (DO) prior to and following a 5 days period at 20°C. A known volume of sample of waste water is diluted with a known volume of dilution water (water containing nutrients for a bacterial growth). The DO of this solution is predetermined (D_1). Now, this solution is kept for incubation for a period of 5 days and DO of this solution is determined (D_2):

$$\text{BOD} = D_1 - D_2 \times \frac{\text{Volume of sample after dilution}}{\text{Volume of sample before dilution}}$$

D_1 = DO of diluted water sample after its preparation

D_2 = DO of diluted water sample after incubation of 5 days at 20°C.

Procedure

Dilution water is prepared in the following manner. Take 1 litre of distilled water in a suitable bottle and add 1 ml of phosphate buffer (pH 7.2), 1 ml of $MgSO_4$ (22.5 g/L), 1 ml of $CaCl_2$ (27.5 g/L) and 1 ml of $FeCl_3$ (0.25 g/L). The dilution depends on the nature of the sample, water is aerated by passing compressed air through a diffusion tube until it is completely saturated.

Take about 5 ml of the waste water sample and dilute it to 300 ml with dilution water as prepared above. Take the diluted sample in two BOD bottles upto the neck. The dissolved oxygen in one bottle is determined immediately and in the other bottle after five days of incubation. The procedure to determine. DO is same as given in experiment 4.4.

Tabulation

Determination of DO in BOD Bottle No. 1

No. of observations	*Volume of the solution taken in the conical flask (ml)*	*I.B.R. (ml)*	*F.B.R. (ml)*	*Volume of hypo solution used (Difference D_1 ml)*
1	100			
2	100			
3	100			

Determination of DO in BOD bottle No. 2 (After Incubation for 5 Days)

No. of observations	*Volume of the solution taken in the conical flask (ml)*	*I.B.R. (ml)*	*F.B.R. (ml)*	*Volume of hypo solution used (Difference D_2 ml)*
1	100			
2	100			
3	100			

Calculation

Volume of sample before dilution = 5 ml

Volume of sample after dilution = 300 ml

$$\text{BOD} = (D_1 - D_2) \times \frac{\text{Volume of sample after dilution}}{\text{Volume of sample before dilution}} \text{ mg/L}$$

Conclusion

BOD of the given waste water sample = ... ppm.

Preparation of Reagents

Prepare the reagents as in case of determination of dissolved oxygen in Experiment No. 4.4.

1. *Phosphate buffer (pH 7.2):* Dissolve 8.50 g $KH_2PO_4.7H_2O$ and 21.75 g Na_2HPO_4 and 1.70 g NH_4Cl in boiled out distilled water and make the volume to 1 litre in a volumetric flask.

EXPERIMENT 4.6

Aim of the Experiment

Determination of chemical oxygen demand in a given waste water sample.

Apparatus Required

- Burette
- Pipette
- Conical flask
- Beaker
- Round bottomed flask (500 ml) with ground glass
- Joint
- Condenser
- Water bath.

Chemicals Required

- $\frac{N}{4}$ $K_2Cr_2O_7$ solution
- $\frac{N}{4}$ Mohr's salt solution
- Ferroin indicator
- $HgSO_4$
- Silver sulphate reagent.

Theory

The chemical oxygen demand is defined as the amount of oxygen used while oxidising the organic matter content of a sample of water with a strong chemical oxidant under acidic conditions. Chemical Oxygen Demand (COD) is a measure of any kind of oxidisable impurities present in the sewage. COD is a parameter for stream and industrial waste water analysis.

Most types of organic matter are completely oxidised by a boiling mixture of chromic acid and sulphuric acid to produce CO_2 and H_2O. A measured quantity of the sample is refluxed with a known amount of $K_2Cr_2O_7$ and H_2SO_4 and the excess of dichromate remaining unreacted is titrated with Mohr's salt solution using ferroin as indicator. The $K_2Cr_2O_7$ consumed is proportional to the amount of oxidisable organic matter measured as oxygen equivalent.

Procedure

Take 50 ml of the given waste water sample in a round bottomed flask and add 1g of $HgSO_4$ and some broken porcelain pieces. Immerse the flask in cold water and slowly add 75 ml silver sulphate reagent with continuous shaking. Now, add 25 ml of 0.25 N $K_2Cr_2O_7$ solution and mix the contents of the flask thoroughly. Attach the reflux condenser and reflux for two hours. Wash the condenser with distilled water into the flask. Cool and dilute to about 300 ml by distilled water. Add 2–3 drops of ferroin as indicator and titrate against 0.25 N Mohr's salt solution till the end point from blue to wine red is reached. Record the volume of Mohr's salt solution used (V_1 ml). Repeat the experiment to get concordant readings.

Perform a blank titration using distilled water in place of sample water. Take 50 ml of distilled water in a round bottomed flask, add the same amounts of reagents and reflux for two hours. Titrate it in the same way as done with the sample water. Record the volume of Mohr's salt solution used (V_2 ml). Repeat the experiment to get concordant readings.

COD of the sample corresponds to $(V_2 - V_1)$ ml of standard mohr's salt solution.

Tabulation

Sample Solution

No. of observations	*Volume of the solution taken in the conical flask (ml)*	*I.B.R. (ml)*	*F.B.R. (ml)*	*Volume of Mohr's salt solution used (Difference V_1 ml)*
1	50			
2	50			
3	50			

Blanck Titration

No. of observations	*Volume of the solution taken in the conical flask (ml)*	*I.B.R. (ml)*	*F.B.R. (ml)*	*Volume of Mohr's salt solution used (Difference V_2 ml)*
1				
2				
3				

Calculation

Volume of the test sample taken = 50 ml

Volume of Mohr's salt solution used in case of sample solution = V_1 ml

Volume of Mohr's salt solution used in case of distilled water = V_2 ml

Normality of Mohr's salt solution = 0.25 N

COD determination = $(V_2 - V_1)$ ml.

$$\underset{\text{(Sample)}}{N_1V_1} = \underset{\text{(Mohr's salt solution)}}{N_2V_2}$$

$$N_1 \times 50 = 0.25\text{ N} \times (V_2 - V_1)$$

or,
$$N_1 = \frac{(V_2 - V_1) \times 0.25\text{ N}}{50}$$

∴
$$\text{COD} = \frac{(V_2 - V_1)}{50} \times 0.25 \times 8\text{ g/L}$$

$$= \frac{V_2 - V_1}{50} \times 0.25 \times 8 \times 1000 \quad \text{mg/L}$$

$$= (V_2 - V_1) \times 40 \quad \text{mg/L.}$$

Conclusion

The COD of the given waste water sample is $(V_2 - V_1) \times 40$ ppm.

Preparation of Reagents

1. *Potassium Dichromate Solution (0.25 N):* Dissolve 12.259 g of dry $K_2Cr_2O_7$ (eq. wt. 49.035) in boiled out distilled water and dilute to 1 L.
2. *Mohr's Salt Solution (0.25 N):* Weigh 98 g of Mohr's salt and transfer in a flask having a few drops of conc. H_2SO_4. Add distilled water to dissolve it and make up the volume to 1 L.
3. *Silver Sulphate Reagent:* Dissolve 5 g of Ag_2SO_4 in 500 ml of conc. H_2SO_4.
4. *Ferroin Indicator:* Dissolve 1.485 g of O-phenanthroline monohydrate with 0.695 g of $FeSO_4.7H_2O$ in distilled water and dilute to 100 ml.

EXPERIMENT 4.7

Aim of the Experiment

Determination of free chlorine in a given water sample.

Apparatus Required

- Burette
- Pipette
- Conical flask
- Glazed tile.

Chemicals Required

- Standard N/20 Hypo solution ($Na_2S_2O_3$)
- Potassium iodide solution (10%)
- Conc. HCl
- Glacial acetic acid
- Starch solution (Freshly prepared).

Theory

Water contains some disease causing bacterias (pathogenic). The process of destroying/ killing the disease causing bacterias, microorganisms from the water and making it fit for use is called disinfection. Chlorine is widely used for disinfection of potable and municipal water supplied to remove the pathogenic organisms. Chlorination is usually done with the help of bleaching powder or chlorine gas or chlorine dissolved in water in the form of concentrated solution or with chloramine. Chlorine (gas/solution) on reaction with water produces hypochlorous acid, which acts as a germicide.

$$Cl_2 + H_2O \longrightarrow HOCl + HCl$$

$$HOCl + \text{Germs} \longrightarrow \text{Germs are killed}$$

However, excess of free chlorine in drinking water is undesirable, as it imparts chlorinous taste to water. Excess of free chlorine in water is also injurious for human metabolism. Hence it is necessary to estimate free chlorine in municipal water prior to the domestic supply so as to make necessary adjustment in dose rate accordingly.

The principle involved in the estimation of free chlorine in water is that when a measured quantity of water is treated with excess of potassium iodide, the free chlorine present in the water oxidises the corresponding amount of potassium iodide to iodine. The liberated iodine is estimated by titrating against standard hypo solution using starch as indicator.

$$Cl_2 + 2KI \longrightarrow 2KCl + I_2$$

$$I_2 + 2Na_2S_2O_3 \longrightarrow Na_2S_4O_6 + 2NaI$$

$$I_2 + \text{Starch} \longrightarrow \text{Deep blue complex}$$

Procedure

Pipette out 100 ml of a given water sample in a conical flask and add 10 ml of KI solution followed by 3 ml of glacial acetic acid to maintain pH upto 3–4. Cover the flask with stopper and mix the solution by shaking the flask. Then wash the sides of the flask using a wash bottle with distilled water. Titrate it with standard $Na_2S_2O_3$ solution taken in the burette till the solution becomes straw yellow. Add 2 ml of starch solution when the solution will turn deep blue. Continue the titration with $Na_2S_2O_3$ till the blue colour disappears. Note down volume of hypo solution used (V_2 ml). Repeat the experiment to get concordant readings.

Tabulation

No. of observations	*Volume of sample taken in conical flask (V_1 ml)*	*I.B.R. (ml)*	*F.B.R. (ml)*	*Volume of hypo solution used (Difference V_2 ml)*
1	100			
2	100			
3	100			

Calculation

N_1 = Normality of sample with respect to free chlorine = ?

V_1 = Volume of the sample = 100 ml

N_2 = Normality of hypo used = $\frac{N}{20}$

V_2 = Volume of hypo used = x ml (say)

$$N_1V_1 = N_2V_2$$

$$N_1 \times 100 = \frac{N}{20} \times x$$

$$N_1 = \frac{N}{2000} \times x$$

$$\text{Total chlorine residual} = N_1 \times \text{Eq. wt.}$$

$$= \frac{x}{2000} \times 35.5 \times 1000 \text{ mg/L}$$

$$= \frac{x}{2} \times 35.5 \text{ ppm.}$$

Conclusion

Amount of total free/residual chlorine in given water sample = $\frac{x}{2} \times 35.5$ ppm.

Preparation of Reagents

1. *Sodium Thiosulphate Solution* $\left(\frac{N}{20}\right)$: Dissolve 12.41 g of $Na_2S_2O_3.5H_2O$ in 1 L of distilled water. Standardise with $\frac{N}{20}$ $K_2Cr_2O_7$ to know the exact strength.
2. *Potassium Iodide (10%):* Dissolve 10 g of KI in 100 ml of distilled water.
3. *Starch Indicator (1%):* Prepare a paste of 1 g of powdered starch with distilled water and add to this paste 100 ml of boiled water with constant stirring. Boil for about five minutes and then cool.

EXPERIMENT 4.8

Aim of the Experiment

Determination of chloride ion in a given water sample by Argentometric method (Mohr's method).

Apparatus Required

- Burette
- Pipette
- Conical flask
- Beaker.

Chemicals Required

- $\frac{N}{50}$ $AgNO_3$ solution
- Potassium chromate indicator solution (K_2CrO_4)
- Calcium carbonate
- Methyl orange indicator.

Theory

Chlorides are generally present in water in the form of the NaCl and $MgCl_2$ and may be due to leaching of marine sedimentary deposits. Chloride concentration over 250 ppm imparts a peculiar taste to water and is therefore objectionable. Presence of chlorides in the boiler water leads to problems like priming and foaming. The chloride concentration of raw water should therefore regularly be tested. Chloride ions can be determined by titration with standard silver nitrate solution using potassium chromate as indicator (Mohr's method). As the titration proceeds, the chloride ions present react with $AgNO_3$ forming insoluble white precipitate of AgCl. As soon as all the chlorides are precipitated out, the extra one drop of $AgNO_3$ reacts with the indicator forming red silver chromate. This indicates the end point

$$AgNO_3 + NaCl \longrightarrow AgCl\downarrow + NaNO_3$$

$$AgNO_3 + K_2CrO_4 \longrightarrow Ag_2CrO_4 + 2KNO_3$$

It should be noted here that the water sample used for chloride determination should be practically neutral or faintly alkaline (pH 7–8) since silver chromate is readily soluble in dilute acids, whereas in strongly alkaline solutions, Ag_2O may get precipitated.

The required pH range can be achieved easily by adding a pinch of pure $CaCO_3$. Excess $CaCO_3$, being insoluble does not interfere.

Procedure

1. *Blank Correction:* Transfer 50 ml of distilled water in a conical flask and add 3–4 drops of K_2CrO_4 indicator. Titrate it against N/50 $AgNO_3$ solution taken in the burette till the brick red colour appears in the solution. Note down the reading (X ml) and repeat to get concordant readings.

   This volume has to be deducted from the volume of the $AgNO_3$ solution used in titrating the sample water.
2. *Sample Titration:* Transfer 50 ml of given water sample and add a drop of methyl orange indicator. If red or pink colour appears, add a pinch of $CaCO_3$, so that red or pink colour disappears. Add 1 ml of K_2CrO_4 indicator solution and titrate against $\frac{N}{50}$ $AgNO_3$ solution

taken in the burette until the end point, when red colour persists. Repeat the titration to get concordant readings. Note down the titre value (Y ml).

Tabulation

Blank Correction

No. of observations	*Volume of the sample taken in the conical flask (ml)*	*I.B.R. (ml)*	*F.B.R. (ml)*	*Volume of $AgNO_3$ solution used (Difference X ml)*
1	50			
2	50			
3	50			

Sample Titration

No. of observations	*Volume of the sample taken in the conical flask (ml)*	*I.B.R. (ml)*	*F.B.R. (ml)*	*Volume of $AgNO_3$ solution used (Difference Y ml)*
1	50			
2	50			
3	50			

Calculation

Normality of the sample with respect to chloride ion = N_1 = ?

Volume of the sample water = $V_1 = 50$

Normality of $AgNO_3$ solution used = N/50

Volume of the $AgNO_3$ solution used = $V_2 = (Y - X)$ ml

$$N_1V_1 = N_2V_2$$

$$N_1 \times 50 = \frac{N}{50} \times (Y - X)$$

$$N_1 = \frac{N}{50 \times 50} \times (Y - X)$$

$$\text{Strength of } Cl^- \text{ ions} = N_1 \times \text{Eq. wt. of } Cl^-$$

$$= \frac{Y - X}{50 \times 50} \times 35.5 \text{ g/L}$$

$$= \frac{Y - X}{50 \times 50} \times 35.5 \times 1000 \text{ mg/L.}$$

Conclusion

The chloride ion concentration in the given water sample is

$$\frac{Y - X}{50 \times 50} \times 35.5 \times 1000 \text{ ppm.}$$

Preparation of Reagents

1. *Silver Nitrate Solution (N/50):* Dissolve 3.398 g of $AgNO_3$ in distilled water and make up the volume to 1 L.
2. *Potassium Chromate Indicator (K_2CrO_4):* Dissolve 5 g of potassium chromate in 100 ml of distilled water. Use 1 ml of the indicator for each titration.
3. *Methyl Orange Indicator:* Dissolve 50 mg of free acid form of methyl orange solid in distilled water and make the volume to 100 ml with distilled water. Filter the solution if necessary.

VIVA-VOCE

1. What do you mean by temporary and permanent hardness?
2. What is difference between hard water and heavy water?
3. What is the principle of EDTA titration?
4. How temporary hardness of water can be removed?
5. What are the disadvantages of hard water?
6. Why the pH of the solution is maintained 10 in determination of hardness by EDTA method?

   Ans. At higher pH values, $CaCO_3$ or $Mg(OH)_2$ may get precipitated and indicator may change its colour. At lower pH values the Mg-indicator complex becomes unstable and a sharp end point can be obtained.
7. Why hardness is generally expressed as calcium carbonate equivalent?

   Ans. Hardness of water is generally expressed as calcium carbonate equivalent because it is most insoluble salt and its molecular weight is 100, which makes the calculation easier.
8. What is the source of alkalinity in water?
9. What is alkalimetry?

   Ans. The method used to determine the strength of a base by titrating against a standard acid solution using suitable indicator is known as alkalimetry.
10. What are the disadvantages of highly alkaline water?
11. What is the significant of determination of dissolved oxygen in water?
12. Define the term BOD.
13. What is the significance of BOD determination?
14. What is BOD5 test?
15. Define COD.

16. What is the structure of ferroin indicator?

Ans. Ferroin is a ferrous complex of the organic compound 1, 10-phenanthroline.

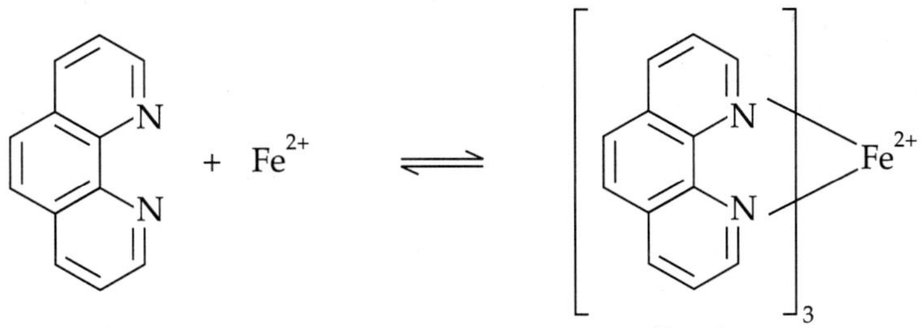

Ferroin

17. What is the significance of COD determination?

18. Why chlorine is added to water?

19. What is the action of chlorine on pathogens?

20. Why and how pH is adjusted between 3 and 4 in the determination of free chlorine?

Ans. pH is adjusted between 3–4 using acetic acid. At neutral pH some combined chlorine residuals do not liberate I_2 by reacting with KI.

21. In which form the chlorides are present in water.

Ans. Chlorides are present in water usually as NaCl, $MgCl_2$ and $CaCl_2$

22. What is the principle of argentometric titrations?

23. Why a pH of 7 to 8 is maintained in chlorine estimation in water?

Ans. Water sample used for chloride determination should be neutral or faintly alkaline, because silver chromate is readily soluble in dilute acids and strongly alkaline Ag_2O may get precipitated.

Chapter 5 FUELS

Fuel is a combustible substance containing carbon as the major constituent which on proper burning gives large amount of heat that can be used economically for domestic and industrial purposes. The classification of fuels is according to occurrence and state of aggregation (physical state). According to occurrence fuel is of natural (primary) or derived (secondary) type and according to state of aggregation the fuels are classified as solid, liquid and gaseous fuels.

EXPERIMENT 5.1

Aim of the Experiment

Determination of moisture, volatile and ash contents in a given sample of coal by proximate analysis.

Apparatus Required

- Analytical balance
- Electric oven
- Muffle furnace
- Desiccator
- Silica crucible with lid.

Materials

- Given coal sample (powdered).

Theory

Proximate analysis is the simplest type of analysis of coal which informs its practical utility. It includes determination of moisture, volatile matter, ash and fixed carbon.

During the burning of coal moisture escapes out as water vapour. It requires latent heat of vapourisation. As a result less heat is liberated for which the calorific value of coal is lowered. Presence of water vapour also quenches the fire ; as the supply of oxygen is reduced. However if moisture is present upto 10% in coal, the fuel used becomes uniform and there will be less flyash, which would have otherwise choked the chimneys.

Ash is an useless non-combustible matter present in coal. It reduces the calorific value of coal. It also causes hindrance to flow of air and heat into burning sample for which the temperature is reduced in the furnace. It causes trouble during firing of furnace by forming clinkers (fused ash lumps). As a result, the air passage becomes blocked and burning of coal becomes irregular. Thus lower the ash content, better is the quality of coal. The presence of more ash increases the cost of transportation, handling and storage. It causes wear and tear of furnace walls and feeding accessories.

The volatile matter is not a constituent of coal but consists of a complex mixture of gaseous and liquid products resulting from the thermal decomposition of coal substance. The volatile matter of coal may include combustible gases like CO, H_2, CH_4 etc., and non combustible gases like N_2, CO_2 etc. The coal containing high volatile matter burns with a long flame, high smoke for which its calorific value becomes less. The volatile matter percentage gives some idea about coking property of the coal.

Fixed carbon content in coal should be higher so that it has more calorific value. Fixed carbon amount in coal varies inversely with volatile matter.

Procedure

The method of analysis of coal are selected from those of the United States Bureau of mines and of the American Society for Testing Materials (ASTM).

1. *Determination of Moisture:* Take 1 g of air dried given coal sample (powdered) in a previously weighed dry porcelain crucible by an analytical balance. Keep it inside the electric oven maintained at 105°C for an hour. Then remove the crucible with a pair of tongs and cool inside a desiccator. Then accurately weigh on analytical balance. The difference in weight is reported as percentage moisture in the coal sample.
2. *Determination of Volatile Matter:* Take the dried sample of coal after moisture analysis and heat in a muffle furnace at 925 ± 20°C for exactly seven minutes. During heating the crucible containing the sample should be covered with a lid. Remove the crucible from the muffle furnace with the help of long legged tong carefully. Cool it outside then inside a desiccator. Then accurately weigh it. Loss in weight is reported as percentage of volatile matter.
3. *Determination of Ash:* Heat the residual coal in the crucible after step 2 inside a muffle furnace maintained at 725 ± 25°C for nearly half hour. Occasionally stir the ash with a platinum wire. Then cool the crucible first in air and then in desiccator and take the weight. Repeat the heating until a constant weight of crucible is obtained.
4. *Fixed Carbon:* The sum total of percentages of volatile matter, moisture and ash subtracted from 100 gives the percentage of fixed carbon.

Observation and Calculation

1. *Moisture:*

Weight of empty crucible = W_1 g

Weight of crucible + coal sample = W_2 g

Weight of crucible + coal sample after heating = W_3 g

$$\% \text{ Moisture} = \frac{W_2 - W_3}{W_2 - W_1} \times 100 = X$$

2. *Volatile Matter:*

Weight of empty crucible = W_1 g

Weight of crucible + coal sample = W_2 g

Weight of crucible + coal sample after heating = W_4 g

$$\% \text{ Volatile matter} = \frac{W_3 - W_4}{W_2 - W_1} \times 100 = Y$$

3. *Ash:*

Weight of empty crucible = W_1 g

Weight of crucible + coal sample = W_2 g

Weight of crucible + ash = W_5 g.

$$\% \text{ Ash} = \frac{W_5 - W_1}{W_2 - W_1} \times 100 = Z$$

4. *Fixed Carbon:*

$$\% \text{ Fixed carbon} = 100 - (X + Y + Z) = V.$$

Conclusion

The given coal sample after proximate analysis contains

Moisture = X%

Volatile matter = Y%

Ash = Z%

Fixed carbon = V%.

EXPERIMENT 5.2

Aim of the Experiment

Flue gas analysis by Orsat's apparatus.

Apparatus Required

- Orsat apparatus.

Chemicals Required

- Potassium hydroxide solution
- Alkaline pyrogallol solution
- Ammoniacal cuprous chloride solution.

Theory

Gas analysis is a branch of volumetric analysis. Gases are more conveniently measured by volume than by weight. Gases may be met with and determined singly or in mixtures. The best example utilised in general gas analysis is the analysis of flue gases. Normally these contain carbon monoxide, carbon dioxide, oxygen, hydrogen, methane, ethane, nitrogen and unsaturated hydrocarbons like ethylene, propylene and benzene.

Typical methods used in volumetric gas analysis are,

1. Burning the gas with air or oxygen and measuring the change in volume and the amount of CO_2 produced.
2. Successive absorption of gases by suitable reagents each of which absorbs one gas only.

An idea about the complete or incomplete combustion of fuel is made by the analysis of flue gas. If the flue gas contains excessive CO, it indicates incomplete combustion, thereby considerable wastage of fuel. It also indicates that supply of oxygen is poor.

The analysis of flue gases is done with the help of Orsat's apparatus (Fig. 5.1).

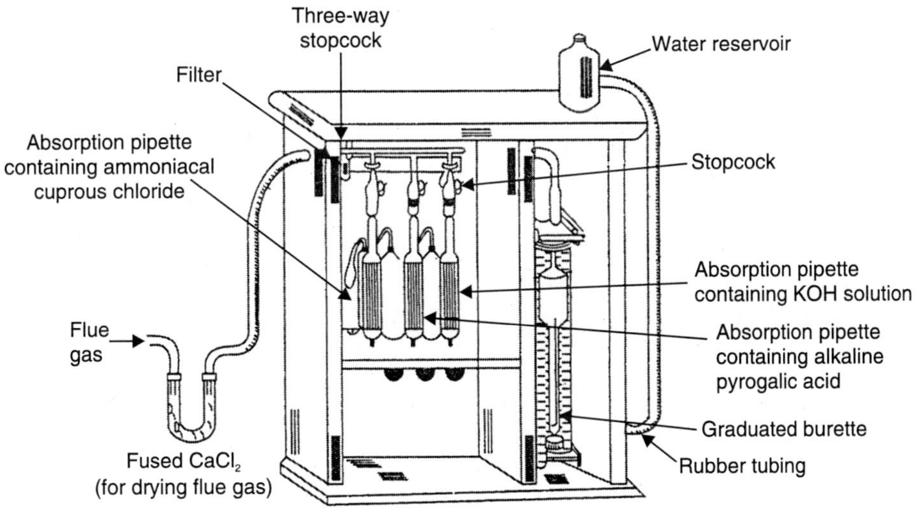

FIGURE 5.1 **Orsat's gas analysis apparatus.**

Description of the Apparatus

The simplest type of Orsat apparatus contains three absorption pipettes, connected to a water jacketed measuring burette. The graduated burette is surrounded by a water jacket to keep the temperature of gas constant during the experiment. The lower end of the burette is connected to a water reservoir by means of a long rubber tubing. The absorption bulbs are usually filled with glass tubes, so that the surface area of contact between the gas and the solution is increased.

The absorption bulbs have solutions for the absorption of CO_2, O_2 and CO respectively. First bulb has potassium hydroxide solution and it absorbs only CO_2. The second bulb has a solution of alkaline pyrogallol solution and it can absorb CO_2 and O_2. The third bulb contains ammoniacal cuprous chloride and it can absorb CO_2, O_2 and CO. Hence it is necessary that the flue gas is passed first through potassium hydroxide bulb where CO_2 is absorbed, then through alkaline pyrogallol solution bulb when only O_2 will be absorbed (as CO_2 has already been removed) and finally through ammoniacal cuprous chloride bulb, where only CO will be absorbed.

Procedure

Clean the whole apparatus, and make it air tight. Fill the absorption bulbs with respective solutions and close the stop cocks. Fill the jackets and levelling reservoir with water. Now open the 3 way stopcock and raise the reservoir till the burette is completely filled with water and air is excluded from the burette. Connect the three way stopcock flue gas supply and lower the reservoir to draw the gas to be analysed. Again open the three way stopcock to the atmosphere and raise the reservoir so that gas is expelled out. Repeat this sucking and expelling the gas thrice. Now suck the gas into burette and adjust the volume to 100 ml at 1 atmospheric pressure.

Open the stopper of the bulb containing KOH and raise the reservoir so that gas is forced into bulb. Again send the gas to the burette. Repeat this seven times for complete absorption of CO_2 in KOH. Take the unabsorbed gas to the burette by ensuring that the level of solution in CO_2 absorption bulb stands at a fixed mark. Now close the stopcock and note the volume of the residual gas by equalising the levels of water in burette and reservoir.

The decrease in volume gives the volume of residual gas. The decrease in volume gives the volume of CO_2 in 100 ml gas sample.

Determine the volumes of CO and oxygen by passing the residual gas through ammoniacal cuprous chloride bulb and alkaline pyrogallol bulb respectively. The gas remaining after absorption of CO_2, CO and oxygen are taken to be nitrogen.

Always take precaution to pass the gas through KOH solution first, alkaline pyrogallol second and ammoniacal cuprous chloride last. Because alkaline pyrogallol can also absorb some CO_2 so that the percentage of oxygen will become more.

Repeat the experiment with fresh sample to obtain concordant readings.

Observation

Volume of the flue gas sample = 100 ml

$$CO_2 = X \text{ ml}$$
$$O_2 = Y \text{ ml}$$
$$CO = Z \text{ ml.}$$

Conclusion

The flue gas analysis is

$$CO_2 = X\%$$
$$O_2 = Y\%$$
$$CO = Z\%$$
$$\% N_2 = 100 - (X + Y + Z)$$

Preparation of Reagents

1. *Potassium Hydroxide Solution:* Dissolve 450 g of stick potash in 500 ml of freshly boiled distilled water, cool before use.
2. *Alkaline Pyrogallol Solution:* Dissolve 30 g of pure pyrogallic acid in 100 ml of distilled water. Add 200 ml of the above caustic potash solution. Avoid contact with air while preparing the solution.
3. *Ammoniacal Cuprous Chloride Solution:* Add 75 g of white cuprous chloride to a solution of 15 g of ammonium chloride in 80 ml water, and then add sufficient conc. aqueous ammonia until solution is complete. Avoid large excess of ammonia to be used. Keep the solution in contact with copper, gauze, wire or turnings in a stoppered bottle.

VIVA-VOCE

1. What are the fuels and how they are classified?
2. What is proximate analysis?
3. What is calorific value of a fuel?

   Ans. Calorific value is defined as the total quantity of heat liberated by the complete combustion of one unit of fuel in oxygen.
4. What is the calorific value of anthracite coal?

   Ans. The calorific value of anthracite coal is 8,650–8,700 KCal/kg.
5. What do you mean by flue gas?

Chapter 6 LUBRICANTS

A substance capable of reducing the friction between the two surfaces which are sliding over each other is called lubricant. In lubrication, lubricant is used to reduce the coefficient of friction between two sliding surfaces of machinery. The lubricant does not allow the direct contact between the sliding surfaces by keeping its own film in between these sliding surfaces. Lubricant also acts as a coolant or heat transfer medium by removing the frictional heat generated due to the sliding surfaces. Lubricant acts as a seal in internal combustion engines by sealing the piston and cylinder wall at the compression rings so that there is no leakage of gases at high pressure in the combustion chamber and in the way power loss is averted by lubrication in internal combustion engines.

The lubricating oils originally used were fatty oils of animal or vegetable origin such as lard oil, tallow oil, castor oil, rape seed oil etc. But with the advent of petroleum industry, these have been replaced by mineral oils. However, the fatty oils are still used as additives to petroleum oils and for the manufacture of grease.

The selection of lubricant for a particular job can be done on the basis of various tests based on their properties. The following tests are usually conducted for the preliminary assessment of a lubricant for its suitability for a particular use.

- Viscosity
- Flash-point and fire-point
- Acid value
- Iodine value.

EXPERIMENT 6.1

Aim of the Experiment

Determination of viscosity of an oil by redwood viscometer.

Apparatus Required

- Red wood viscometer
- Thermometer
- Stopwatch
- Kohlrousch flask.

Materials

- Given sample of lubricating oil.

Theory

Viscosity is one of the most important properties of any lubricating oil. This indicates us about the suitability of the oil for lubricating purposes. A lubricant must reduce friction between sliding parts of any machine. This avoids the direct metal to metal contact. The main criteria of a lubricant is that it should be sufficiently viscous under high temperature and pressure exerted by the machine to adhere to the surface. If the viscosity is low then a thin film of lubricant can not adhere to the sliding surfaces. In case the viscosity is high there will be excessive friction. The absolute viscosity of a fluid oil can be determined by measuring the rate of flow of the oil through a capillary tube kept at a uniform temperature. But in case of lubricating oil specific viscosity is generally determined by measuring the time taken for a given quantity of oil to flow through an orifice or jet of standard dimension under standard conditions.

Measurement of viscosity of lubricating oil is made with the help of an apparatus called Red wood viscometer. Redwood viscometer is commonly used for determining. Viscosities of thin lubricating oils. There are two types of Red wood viscometer *i.e.,* Red wood viscometer No. 1 and Red wood viscometer No. 2. Red wood viscometer No. 1 will correctly indicate the viscosity for a liquid having time flow between 30 sec to 2000 sec. If the time flow measured with this apparatus for any oil exceeds 2000 seconds, the test should be repeated with Red wood viscometer No. 2 which will give the correct value of viscosity for such highly viscous oil.

Description of the apparatus

Red wood viscometer consists of the following parts (Fig. 6.1).

1. *Oil Cup:* It is a 90 mm in height and 46.5 mm in diameter silver plated brass cylinder. Its upper end is open. Its lower end is fitted with an agate jet having bore of diameter 1.62 mm and length 10 mm. The jet can be opened and closed by a valve rod. The valve rod is a small silver plated brass ball fixed to a stout wire. There is a pointer to indicate the level to which the cylinder is to be filled with oil. The pointer is fixed on inner side of the cylinder. The cover of the cup is filled with a thermometer to indicate the temperature of oil.
2. *Heating Bath:* There is a cylindrical copper bath which surrounds the oil cup. This copper bath contains water. It is provided with an outlet tap to let out water from it and a long side tube projecting outwards. There is a thermometer to indicate the temperature of water.
3. *Stirrer:* The heating bath is provided with a stirrer which stires the water in the heating bath for maintaining uniform desired temperature. The stirrer is sealed at the top to prevent water rushing into the oil cylinder.
4. *Spirit Level:* The cover of the cup is provided with a spirit level for vertical levelling of the jet.

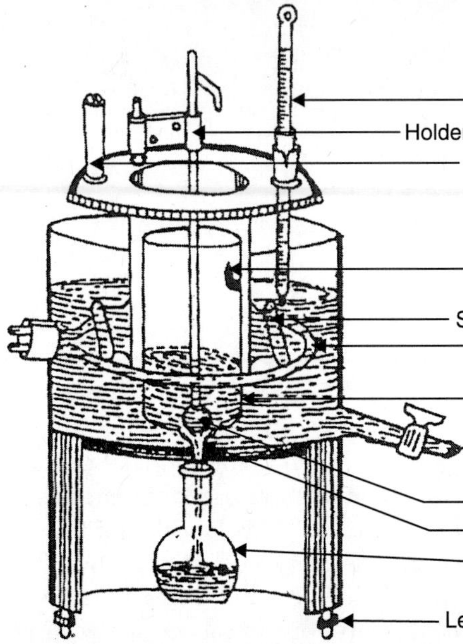

FIGURE 6.1 **Cross-section of redwood viscometer.**

5. *Levelling Screws:* The entire apparatus rests on three legs provided at the bottom with levelling screws.
6. *Kohlarausch Flask:* This flask receives the oil from jet outlet. Its capacity is 50 ml upto the mark in neck.

Procedure

Accurately level the entire apparatus. Clean the oil cap with the help of a suitable solvent and properly dry to remove any trace of the solvent. Fill the bath with water to determine viscosity at 80°C and below. For higher temperature fill the bath with an oil having suitable viscosity at the test temperature. Keep the brass ball in position so as to seal the orifice. Now pour the supplied oil sample under test carefully into the oil cup up to the mark. Keep the kohlarausch flask in the position below the jet. Insert a thermometer and a stirrer and cover the lid. Keep stirring the water in the bath and oil in the oil cup and adjust the bath temperature until the oil attains the desired constant temperature. When the temperature of the oil has become quite steady in the oil cup and shows a constant reading. Lift the ball valves and simultaneously start the stop watch, allow the oil to fill in the flask upto 50 ml mark. Stop the stopwatch and note down the time in seconds. Replace the ball valve in position to seal the cup to prevent overflow of the oil. Again refill the oil cup upto the mark and repeat the experiment to get nearly reproducible results. Repeat the experiment at five elevated temperatures say 40°C, 50°C, 60°C, 70°C and 80°C and note the respective time of flow as described below.

Tabulation

No. of observations	*Temperature (°C)*	*Time of flow (sec.) (Red wood viscosity)*	*Kinematic viscosity (Centistokes)*
1	Room Temperature		
2	40		
3	50		
4	60		
5	70		
6	80		

Calculation

The ratio of absolute viscosity to density for any fluid is known as its absolute kinematic viscosity. Since the instruments used are of standard dimensions, kinematic viscosity of the oil in centistokes can be calculated from the time taken by the oil to flow through the standard orifice of the instrument with the help of following equation.

The viscosity of the given oil sample with the help of redwood viscometer at t°C = ... Redwood seconds.

$$V = At - \frac{B}{t}$$

where V = kinematic viscosity of the oil in centistokes

t = Time of flow in seconds

A and B are instrument constants

The value of

A = 0.264 and B = 190; when t = 40 to 85 seconds

A = 0.247 and B = 65; when t = 85 to 2000 seconds.

Conclusion

The given sample of lubricating oil has,

Red wood viscosity at t°C = ... Redwood seconds

Kinematic viscosity at t°C = ... Centistokes.

EXPERIMENT 6.2

Aim of the Experiment

Determination of Flash point and Fire point of a given oil sample by Pensky-Marten's, closed cup apparatus.

Apparatus Required

- Pensky-Marten's closed cup apparatus.

Material

- Given sample of lubricating oil.

Theory

Flash point is defined as the lowest temperature at which it gives off just enough vapours to produce a combustible mixtures which will flash over the entire surface of the sample. After the flash point has been obtained, if the temperature is further raised a point is reached at which the emission of inflammable vapours occurs rapidly enough to support a flame. This is called the Fire point. Fire point temperature is generally higher by 5 to 40°C than that of the flash point. A good lubricating oil has high flash and fire point and thus has high working temperature. This ensures safety against fire hazards during the storage, transport and use of the lubricating oil. If a liquid is having flash point less than 140°F, they are called flammable liquid and those with flash point above 140°F are called combustible liquid. The Pensky-Marten's closed cup apparatus is used for oils with flash point above 120° F.

Description of the Apparatus

1. *Oil Cup:* It is about 5 cm in diameter and 5.5 cm deep. The level to which oil is to be filled is marked inside the cup. The cup lid is provided with 4 openings of standard size. Through one passes a thermometer, through the second, the test flame is introduced, through the third the stirrer carrying 2 brass blades are introduced into the apparatus. The fourth is meant for admission of air.

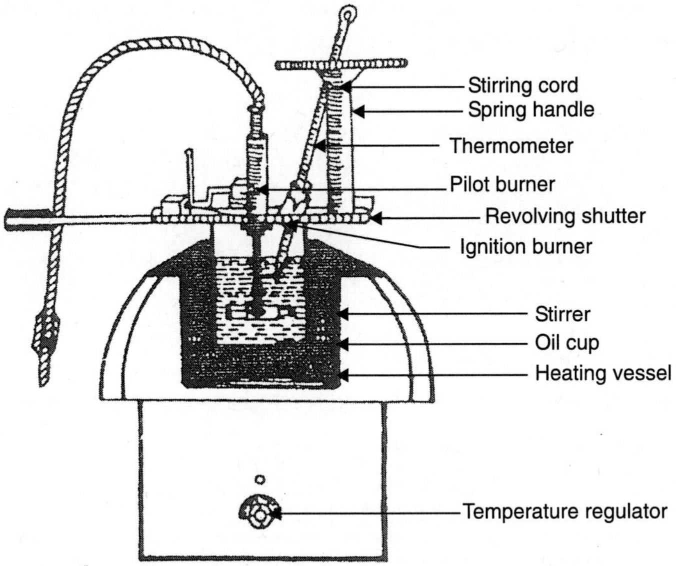

FIGURE 6.2 **Cross-section of pensky-Marten's closed cup apparatus.**

2. *Shutter:* It is a level mechanism provided at the top. By moving the shutter, the opening in the lid opens and flame carried by the flame exposure device is introduced into the opening. Thus the flame can be brought over the oil surface.

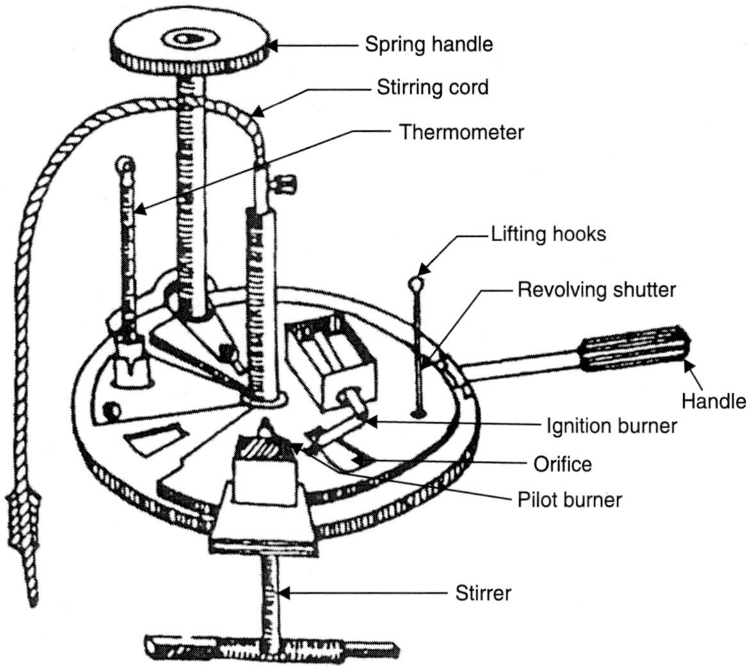

FIGURE 6.3 **Revolving shutter and accessories.**

3. *Flame Exposure Device:* It is a small flame connected to the shutter by a level mechanism.
4. *Air Bath:* Oil cup is supported by its oil flange on an air bath heated by a gas burner.
5. *Pilot Burner:* As the test flame is introduced in the opening, it gets extinguished. But when the test flame is returned to the original position it is automatically lighted by the pilot burner.

Procedure

Thoroughly clean and dry all parts of the cup and its accessories. Take the lubricating oil free from suspended impurities and fill up to the mark in the oil cup and then heat the air bath with a Bunsen burner. Rotate the stirrer in the oil cup at the rate of 2 revolutions per second. Heat the oil bath so as to raise the temperature by 5°C in 1 minute. At every 1°C rise in temperature, introduce the test flame for a moment by moving the shutter. The temperature at which a distinct flash appears inside the cup gives the flash point of oil. At the flash point, there is a combination of weak sound and light. The heating is continued when the oil ignites and continues to burn for 5 seconds. The recorded temperature gives the fire point of the oil.

Tabulation

No. of observations	*Flash point* (°C)	*Fire point* (°C)
1		
2		
3		

Conclusion

The given sample of lubricating oil has

Flash point = ... °F

Fire point = ... °F.

EXPERIMENT 6.3

Aim of the Experiment

Determination of acid value of an oil.

Apparatus Required

- Burette
- Pipette
- Conical flask
- Glazed tile.

Chemicals Required

- $\frac{N}{10}$ KOH solution
- Neutral ethyl alcohol
- Phenolphthalein indicator
- Given sample of lubricating oil.

Theory

The acid value of a lubricating oil is defined as the number of milligrams of potassium hydroxide required to neutralise the free acid present in 1 g of the oil sample.

In good lubricating oils, the acid value should be minimum (< 0.1). Increase in acid value should be taken as an indicator of oxidation of the oil which may lead to gum and sludge formation besides corrosion.

A known weight of the oil sample is dissolved in a suitable solvent and titrated with a standard alcoholic solution of KOH to a definite end point.

$$H^+ + OH^- \longrightarrow H_2O$$

$$C_{17}H_{35}COOH + KOH \longrightarrow C_{17}H_{35}COOK + H_2O$$

Procedure

Weigh out accurately about 5 g of the given oil sample into a conical flask and add 50 ml of neutral alcohol. Heat the flask over a water bath for about 30 minutes. Cool the flask and the contents to room temperature and add a few drops of phenolphthalein indicator. Titrate it with N/10 KOH solution until a faint permanent pink colour appears at the end point. Note down the titre value and repeat the steps to get concordant readings.

Tabulation

No. of observations	*I.B.R. (ml)*	*F.B.R. (ml)*	*Volume of KOH solution used (Difference V ml)*
1 2 3			

Calculation

Weight of the oil sample taken = 5 g

Volume of $\frac{N}{10}$ KOH solution used = V ml

$$\text{Acid value} = \frac{V}{10} \times 5.6 = X \text{ mg (say)}$$

where 5.6 represents the amount of KOH in milligram present per each ml of $\frac{N}{10}$ KOH solution

1000 ml. of 1 N KOH ≡ 56 g of KOH.

Conclusion

The acid value of given oil sample is X mg.

Preparation of Reagents

1. *Potassium Hydroxide Solution (N/10):* Dissolve about 5.6 g of KOH in 1 L of distilled water.
2. *Neutral Ethyl Alcohol:* Neutralise the ethyl alcohol with N/40 NaOH using phenolphthalein as indicator.
3. *Phenolphthalein Indicator:* Dissolve 1 g of phenolphthalein in 100 ml of distilled ethyl alcohol and add 100 ml of distilled water with constant stirring. Filter of necessary.

EXPERIMENT 6.4

Aim of the Experiment

Determination of iodine value of an oil.

Apparatus Required

- Glass stoppered bottle
- Burette
- Pipette
- Conical flask
- Glazed tile

Chemicals Required

- $\frac{N}{10}$ $Na_2S_2O_3$ solution
- Wij's solution
- Potassium iodide solution (10%)
- Chloroform
- Starch solution (freshly prepared).

Theory

Iodine value is a measure of the extent of unsaturation in oils. It is defined as the number of grams of iodine that adds across the carbon-carbon double bonds present in 100 g of oil.

Determination of iodine is done through iodometric titration. To a known weight of oil sample excess iodine monochloride ICl (wij's solution) is added so that some free ICl remains after the addition reaction is complete. Then some KI is added so that iodine is liberated by the reaction between ICl and KI.

$$ICl + KI \longrightarrow KCl + I_2$$

This liberated iodine is estimated by standard hypo solution using starch as indicator

$$2Na_2S_2O_3 + I_2 \longrightarrow Na_2S_4O_6 + 2NaI$$

Procedure

Take two clean and dry 200 ml glass stoppered bottles. Accurately weigh about 5 g of the sample oil under test into one of these two bottles. The other one serves as the blank. Add 5 ml chloroform to both the bottles and shake well. Then add 25 ml of Wij's solution to both the bottles. Allow them to stand for one hour in dark. Then rinse the inside wall and stopper of the bottles. Add 5 ml of 10% kI solution to each bottle and titrate the contents of each bottle separately against N/10 $Na_2S_2O_3$ solution. Add one ml starch solution and continue titration till the blue colour just disappears. Note down the titre values and repeat the experiment to get concordant readings.

Tabulation

Blank Titration

No. of observations	*I.B.R. (ml)*	*F.B.R. (ml)*	*Vol. of $Na_2S_2O_3$ solution used (Difference V_1 ml)*
1			
2			
3			

Sample solution

No. of observations	*I.B.R. (ml)*	*F.B.R. (ml)*	*Vol. of $Na_2S_2O_3$ solution used (Difference V_2 ml)*
1			
2			
3			

Calculation

Let the weight of oil sample taken = W g

$$\text{Strength of } Na_2S_2O_3 \text{ solution} = \frac{N}{10}$$

Volume of thiosulphate required in blank = V_1 ml

Volume of thiosulphate required in case of oil sample = V_2 ml

Volume of thiosulphate which is equivalent to the amount of iodine required for the addition reaction across carbon-carbon double bond in oil sample

$$= (V_1 - V_2) \text{ ml}$$

$$1000 \text{ ml of } 1\text{ N } Na_2S_2O_3 \text{ solution} \equiv 1000 \text{ ml } 1\text{ N } I_2 \text{ solution}$$
$$\equiv 127 \text{ g of } I_2$$

$$\text{So, } (V_1 - V_2) \text{ ml of } \frac{N}{10} Na_2S_2O_3 \text{ solution} \equiv \frac{127\,(V_1 - V_2)}{1000} \times \frac{1}{10} \text{ g of } I_2 = X \text{ g of } I_2$$

Wg of oil add up X g of I_2

$$100 \text{ g of the oil add up } \frac{100\text{ X}}{\text{W}} \text{ g of } I_2$$

$$\text{Hence, Iodine value of given oil sample} = \frac{100\text{ X}}{\text{W}}.$$

Conclusion

$$\text{The iodine value of given oil sample} = \frac{100\text{ X}}{\text{W}}.$$

Preparation of Reagents

1. *Wij's Solution:* Dissolve 8.5 g of iodine and 7.8 g of iodine trichloride in separate portions of about 450 ml. glacial acetic acid. Mix both the solutions and make upto 1 L.
2. *Sodium Thiosulphate (N/10):* Dissolve 24.82 g of $Na_2S_2O_3 \cdot 5H_2O$ (eq. wt. 248.21) in 1 L of boiled out distilled water. Standardise it with $K_2Cr_2O_7$ to know the exact strength.
3. *Potassium Iodide Solution (10%):* Dissolve 10 g of KI in 100 ml of distilled water.

4. *Starch Indicator (1%):* Prepare a paste of 1 g of powdered starch with distilled water and add to this paste 100 ml of boiled water with constant stirring. Boil for about five minutes and then cool.

VIVA-VOCE

1. What are lubricants?
2. What are the unit of viscosity?
3. What is the effect of temperature on viscosity of liquid and gas?
4. Define flash point and fire point of a lubricating oil.
5. What should be the flash point of a good lubricant?

   Ans. Flash point must be atleast above the temperature at which the lubricant is to be used to avoid the risk of fire hazard.
6. What are the factors, that affect the flash and fire point?

   Ans. Moisture, vapour pressure, apparatus used, frequency of application of test flame, rate of heating the test oil etc.
7. How moisture is removed from the lubricating oil?

   Ans. By absorption through dehydrating agents such as Na_2SO_4, anhydrous $CaCl_2$ etc.
8. Define acid value of oil.
9. What is the significance of determination of acid value?
10. What do you mean by iodine value?
11. What is Wij's solution?
12. What is the significance of determination of iodine value?

Chapter 7

ANALYSIS OF ORES AND ALLOYS

The metals are obtained from the natural deposits known as minerals which are found in earth's crust and rocks. A metal bearing mineral from which a metal can be extracted commercially is known as ore. The ores are generally available in the form of oxides, sulphides, carbonates, phosphates, silicates etc. An alloy may be called as a homogenous mixture (solid solution) of two or more metals and also non metals, one of them being a metal. The knowledge about the analysis of ores and alloys is imperative for any engineer to make use of it as an engineering material. Volumetric analysis is widely employed for ores and alloys analysis.

EXPERIMENT 7.1

Aim of the Experiment

Estimation of calcium in limestone.

Apparatus Required

- Burette
- Pipette
- Conical flask
- Volumetric flask
- Funnel
- Beaker.

Chemicals Required

- $\frac{N}{10}$ $KMnO_4$ solution
- Ammonium oxalate (8%)
- Hydrochloric acid (1:1)

- Ammonia
- 5N H_2SO_4
- Ammonium chloride
- Given sample of ore.

Theory

Limestone essentially consists of calcium carbonate but is generally associated with small quantities of magnesium carbonate. The limestone powder is dissolved in HCl and the calcium present in the solution is precipitated as calcium oxalate using ammonium oxalate in presence of ammonia. The precipitate calcium oxalate after washing is treated with 5N H_2SO_4 and the oxalic acid liberated is treated with a standard $KMnO_4$ solution. From the volume of $KMnO_4$ required for the titration, the amount of Ca in the ore can be calculated.

$$\begin{aligned} 1 \text{ ml of } 1 \text{ N } KMnO_4 &\equiv 0.020 \text{ g of Ca} \\ &\equiv 0.028 \text{ g of CaO} \\ &\equiv 0.050 \text{ g of } CaCO_3 \end{aligned}$$

$$CaCO_3 + 2HCl \longrightarrow CaCl_2 + H_2O + CO_2$$

$$CaCl_2 + (NH_4)_2C_2O_4 \longrightarrow CaC_2O_4 + 2NH_4Cl$$

$$CaC_2O_4 + H_2SO_4 \longrightarrow CaSO_4 + H_2C_2O_4$$

$$2KMnO_4 + 3H_2SO_4 \longrightarrow K_2SO_4 + 2MnSO_4 + 5[O] + 3H_2O$$

$$5H_2C_2O_4 + 5[O] \longrightarrow 5H_2O + 10CO_2$$

Procedure

Weigh accurately 1g of the limestone into a beaker and treat with a little amount of 1:1 HCl to dissolve the ore. Take care that there is no loss of the solution due to quick evolution of CO_2 by covering the beaker with a watch glass. After all the ore is dissolved wash the watch glass with distilled water to the same beaker. The solution is to be made alkaline by adding ammonia. Add 1 g of solid NH_4Cl and stir the solution to dissolve it. Continue the stirring and boil and then add 50 ml of 8% ammonium oxalate solution. Boil the solution and allow the calcium oxalate precipitate to settle down. Decant the supernatant liquid through a filter paper. Wash the precipitate several times with hot distilled water containing a little amount of ammonium hydroxide until the filtrate is free from chloride and oxalate ions. Transfer the precipitate from the cone of the filter paper to a beaker with the help of a jet of water. Add about 50 ml hot 5N H_2SO_4 and 40 ml hot distilled water. Warm it with constant stirring. If the precipitate does not dissolve add a little excess 5N H_2SO_4. Cool it and transfer the entire contents of the beaker into a 250 ml volumetric flask. Make up the volume with distilled water and shake well. Pipette out 25 ml of this solution into a conical flask. Add 20 ml 5N H_2SO_4. Heat to about 60–80°C. Titrate the hot solution against standard $KMnO_4$ solution till a faint permanent pink colour is obtained. Repeat the process to get concordant readings and note down the titre value.

Tabulation

No. of observations	*Vol. of the solution taken in conical flask (ml)*	*I.B.R. (ml)*	*F.B.R. (ml)*	*Vol. of $KMnO_4$ used (Difference V ml)*
1	25			
2	25			
3	25			

Calculation

Weight of the ore sample taken = W g

Volume of calcium ions/oxalate solution prepared = 250 ml

Volume of solution taken in conical flask for titration = 25 ml

Normality of $KMnO_4$ solution = $\frac{N}{10}$

Concordant volume of $\frac{N}{10}$ $KMnO_4$ used = V ml

We know that

1 ml of 1N $KMnO_4 \equiv 0.020$ g of Ca

V ml of $\frac{N}{10}$ $KMnO_4 \equiv \frac{0.020 \times V}{10}$ g of Ca

25 ml of the solution contains $\frac{0.020 \times V}{10}$ g of Ca

250 ml of the solution contains $\frac{0.020 \times V}{10} \times 10$ g of Ca = Y g of Ca

Now,

1 g of the ore contains Y g of Ca

100 g of the ore contains Y × 100 g of Ca = M%.

Conclusion

The given limestone ore sample contains M% of Ca.

Preparation of Reagents

1. *Potassium Permanganate Solution* $\left(\frac{N}{10}\right)$: Dissolve 3.1607 g of $KMnO_4$ (eq. wt. 31.607) in distilled water. Boil for 1 hour, cool, filter and make up the volume to 1 L with distilled water. Standardise it with oxalic acid to know the exact strength.
2. *Sulphuric Acid (5N):* Add slowly 139 ml of conc. sulphuric acid (36 N, 98%) to distilled water. After cooling make up the volume to 1 L with distilled water.
3. *Ammonium Oxalate (8%):* Dissolve 8 g of ammonium oxalate in 100 ml of distilled water.
4. *Hydrochloric Acid (1:1):* Mix equal volumes of concentrated HCl and boiled out distilled water.

EXPERIMENT 7.2

Aim of the Experiment

Determination of percentage of copper in brass.

Apparatus Required

- Burette
- Pipette
- Conical flask
- Volumetric flask.

Chemicals Required

- $\frac{N}{10}$ $Na_2S_2O_3$ solution (Hypo)
- Conc. H_2SO_4
- Phosphoric acid
- Ammonia solution
- Potassium iodide
- Starch indicator (Freshly prepared)
- Given brass sample.

Theory

Brass is an alloy of copper and zinc. It may contain some impurities like lead, tin and iron. The composition of brass is as given

Cu (60–80%)	Pb (0–2%)	Fe (0–1%)
Zn (20–40%)	Sn (0–6%).	

The experiment consists of dissolving a known quantity of brass in nitric acid, removing the nitrate by fuming with sulphuric acid adjusting the pH by ammonium hydroxide, complexing the iron present using phosphoric acid and finally titrating the copper ions with hypo by the iodometric method.

When the brass sample is dissolved in nitric acid, the copper present is brought into solution in the form of cupric ions. At the same time, tin is converted into metastannic acid ($SnO_4.4H_2O$), lead and zinc are oxidised to their respective soluble divalent ions, while iron is converted into ferric ions. During the treatment with sulphuric acid nitrate is eliminated, the meta-stannic acid is redissolved and lead ions are precipitated as $PbSO_4$ while all other metals go into solution as their respective sulphates out of all the metal ions present in solution, only Cu^{2+} and Fe^{3+} are reducible by iodide. The interference by iron is eliminated by complexing it with phosphate ion. Then the Cu^{2+} can be determined by the iodometric method.

$$2Cu^{2+} + 4I^- \longrightarrow 2CuI + I_2$$

$$2S_2O_3^- + I_2 \longrightarrow S_4O_6^- + 2I^-$$

$$Cu \equiv I \equiv S_2O_3^-$$

Procedure

Accurately weigh 0.25 g of the given brass sample (air dried) into a conical flask. Add 10 ml of 6N HNO_3 and gently heat the flask until the brass sample is completely dissolved. Then add 10 ml of conc. H_2SO_4 and evaporate the solution on a water bath to white fumes. Allow the mixture to cool and add carefully 20 ml of distilled water slowly from the sides of the flask while cooling under a tap. Boil the solution for 2 minutes and cool. Add liquor ammonia dropwise while shaking the flask vigorously until the appearance of blue cupric ammonium complex. Now add 6 N H_2SO_4 dropwise until the dark blue colour just disappears. Then add 2 ml of phosphoric acid. Cool the solution and transfer to a 100 ml volumetric flask. Make up the volume by using distilled water.

Pipette out 25 ml of this solution into a conical flask and add about 1 g of solid potassium iodide. Shake the flask thoroughly and titrate the solution with standard hypo solution until the solution assumes a faint yellow colour of iodine. Add about 3 ml of freshly prepared starch solution and continue the titration until the blue colour begins to fade. Now add 0.5 g of potassium thiocyanate (KCNS) and continue the titration until the blue colour disappears. Repeat the titration by taking 25 ml of the solution to get concordant readings. Note down the titre value:

Tabulation

No. of observations	*Vol. of solution taken in conical flask (ml)*	*I.B.R. (ml)*	*F.B.R. (ml)*	*Vol. of hypo used (Difference V ml)*
1	25			
2	25			
3	25			

Calculation

Weight of the ore sample taken = W g

Volume of copper solution prepared = 100 ml

Volume of solution taken in conical flask for titration = 25 ml

Normality of hypo solution $= \frac{N}{10}$

Concordant volume of $\frac{N}{10}$ hypo used = V ml

We know that,

1 ml of 1N $Na_2S_2O_3 \equiv 0.06354$ g of Cu

V ml of $\frac{N}{10}$ $Na_2S_2O_3 \equiv \frac{0.06354 \times V}{10}$ g of Cu

25 ml of the solution contains $\frac{0.06354 \times V}{10}$ g of Cu

100 ml of the solution contains $\frac{0.06354 \times V}{10} \times 4$ g of Cu = Y g of Cu

% of Cu in brass = $\frac{Y \times 100}{W}$

Where Y is the amount of copper present in 100 ml of the solution prepared from 'W' g of the brass sample.

Conclusion

Percentage of Cu present in the given brass sample is $\frac{Y \times 100}{W}$.

Preparation of Reagents

1. *Sodium Thiosulphate Solution* $\left(\frac{N}{10}\right)$: Dissolve 24.82 g of $Na_2S_2O_3.5H_2O$ (eq. wt. 248.21) in 1L of boiled out distilled water. Standardise it with $K_2Cr_2O_7$ to know the exact strength.
2. *Starch Indicator (1%):* Prepare a paste of 1 g of powdered starch with distilled water and add to this paste 100 ml of boiled water with constant stirring. Boil for about five minutes and then cool.

EXPERIMENT 7.3

Aim of the Experiment

Determination of chromium in chromite.

Apparatus Required

- Porcelain crucible
- Beaker
- Volumetric flask
- Burette
- Pipette
- Glazed tile.

Chemicals Required

- $\frac{N}{10}$ $K_2Cr_2O_7$ solution
- $\frac{N}{10}$ Ferrous ammonium sulphate solution
- N-phenyl anthranilic acid indicator
- Sodium peroxide.

Theory

Chromite is a highly refractory mineral of chromium. When the ore is fused with excess of sodium peroxide, sodium chromate is formed which goes into solution on leaching the melt with water with simultaneous precipitation of ferric hydroxide.

$$2Fe(CrO_2)_2 + 7Na_2O_2 \longrightarrow 2NaFeO_2 + 4Na_2CrO_4 + 2Na_2O$$
$$NaFeO_2 + 2H_2O \longrightarrow NaOH + Fe(OH)_3$$
$$2Na_2O + 2H_2O \longrightarrow 4NaOH$$

The alkaline solution is boiled off to decompose unreacted sodium peroxide. The solution is filtered, diluted and acidified with HCl. To the dichormate formed a known excess of ferrous ammonium sulphate is added, and the unreacted ferrous ammonium sulphate in the solution is titrated with a standard $K_2Cr_2O_7$ solution using N-phenyl anthranilic acid as indicator.

$$2CrO_4^{2-} + 2H^+ \rightleftharpoons Cr_2O_7^{2-} + H_2O$$
$$Cr_2O_7^{2-} + 6Fe^{2+} + 14H^+ \longrightarrow 2Cr^{3+} + 6Fe^{3+} + 7H_2O$$

$$1 \text{ ml of } \frac{N}{10} K_2Cr_2O_7 \equiv 1 \text{ ml } \frac{N}{10} \text{ Ferrous ammonium sulphate}$$
$$\equiv 0.001733 \text{ g of Cr.}$$

Procedure

Weigh out accurately about 0.5 g of the finely powdered chromite ore into a porcelain crucible. Add 4 g of sodium peroxide and thoroughly mix the two solids with the help of a thin glass rod. Remove any powder adhering to the rod by stirring about 1g of a fresh Na_2O_2 powder with it and transfer this to the crucible in such a way that it covers the ore in the crucible. Replace the lid on the crucible and gently heat the crucible in a fume cupboard over a small flame for about 10 minutes when the charge appears to melt, keep it fused for another 10 minutes at a dull red heat. Allow to cool, and when a solid crust has formed, add 4 g more of Na_2O_2 and fuse the mixture once again at cherry-red heat for another 10 minutes. Allow the crucible to cool and and transfer it to a 60 ml beaker (Pyrex) containing about 25 ml of distilled water. After the violent reaction has subsided, remove the crucible and wash it with distilled water into the same beaker. Boil the liquid for 30 minutes keeping the beaker covered in order to decompose the H_2O_2. Add 250 ml of boiling water and allow the precipitate to settle down. Filter through a sintered glass crucible and wash the residue thoroughly with boiling water until free from chromate.

Evaporate the filtrate to about 200 ml, cool and add cautiously dil. H_2SO_4 until the solution is acidic. Cool and transfer to a 250 ml volumetric flask and make up the volume upto mark with distilled water. Shake well and pipette out 50 ml of this solution into a 500 ml conical flask, add 50 ml of N/10 ferrous ammonium sulphate solution. Add 200 ml of 2N H_2SO_4 and N-phenyl anthranilic acid indicator. Titrate the unreacted ferrous ammonium sulphate solution with N/10 $K_2Cr_2O_7$ taken in the burette until a sharp colour from green to violet red. Repeat the titration to get concordant results.

Standardise the ferrous ammonium sulphate solution against the $\frac{N}{10}$ $K_2Cr_2O_7$ using N-phenyl anthranilic acid indicator. Calculate the volume of ferrous ammonium sulphate solution which was oxidised by the dichromate originating from the 0–5 g of chromite ore taken. From this the amount of chromite present in the chromite ore can be calculated.

Tabulation

No. of observations	*Vol. of the solution taken in conical flask (ml)*	*I.B.R. (ml)*	*F.B.R. (ml)*	*Vol. of $K_2Cr_2O_7$ solution used (Difference V ml)*
1	50			
2	50			
3	50			

Calculation

Wt. of the ore taken = W g

Volume of chromite solution prepared = 250 ml

Volume of the solution taken for titration = 50 ml

Volume of ferrous ammonium sulphate solution added = 50 ml

Normality of ferrous ammonium sulphate solution = N/10

Volume of the $\frac{N}{10}$ $K_2Cr_2O_7$ run down for neutralising the unreacted ferrous ammonium sulphate solution = 30 ml (say)

∴ Volume of $\frac{N}{10}$ ferrous ammonium sulphate consumed by the $Cr_2O_7^{2-}$ present in 50 ml of the solution = (50 – 30) ml = 20 ml

∴ Volume of the $\frac{N}{10}$ ferrous ammonium sulphate consumed by the $Cr_2O_7^{2-}$ in the entire 250 ml of prepared solution = 20 × 5 = 100 ml

1 ml of $\frac{N}{10}$ ferrous ammonium sulphate solution ≡ 0.001733 g of Cr

∴ 100 ml of $\frac{N}{10}$ ferrous ammonium sulphate solution ≡ 0.1733 g of Cr.

So, W g of the chromite ore contains 0.1733 g of Cr

∴ 100 g of the chromite ore contains = $\frac{0.1733 \times 100}{W}$ = Xg of Cr.

Conclusion

Chromium present in the given chromite ore = X%.

Preparation of Reagents

1. *$K_2Cr_2O_7$ Solution (N/10):* Dissolve 4.9035 g of pure and dry $K_2Cr_2O_7$ powder in 1 L of distilled water.

2. *Ferrous Ammonium Sulphate Solution (N/10):* Dissolve 9.8 g of ferrous ammonium sulphate solid in 200 ml of 0.5 M H_2SO_4 in a volumetric flask and make up the volume upto mark to 1 L with boiled and cooled distilled water. Standardise with N/10 $K_2Cr_2O_7$ solution using N-phenyl anthranilic acid as indicator.
3. *N-phenyl Anthranilic Acid Indicator:* Dissolve 0.25 g of solid in 12 ml of N/10 NaOH and dilute to 250 ml with distilled water.

VIVA-VOCE

1. What are the different ores of calcium?

   Ans. Limestone, Dolomite etc.
2. What is difference between ore and mineral?
3. What is the procedure to estimate calcium in dolomite?

   Ans. If the ore is limestone single precipitation is sufficient. In case of dolomite, where a large quantity of magnesium is present, there is every possibility that the precipitate of calcium oxalate may be contaminated with the magnesium oxalate giving higher values of Ca. In order to get the correct values of Ca, dissolve the calcium oxalate precipitate in about 5 ml. of 6N HCl. Make the solution slightly ammonical, dilute with 50 ml. of boiled distilled water and treat again with ammonium oxalate solution (double precipitation is necessary).
4. What is brass and what is its chemical composition?
5. Name some ores of copper.
6. Name some ores of chromium.

   Ans. Chromite, Chrometite.

Chapter 8

DISTRIBUTION LAW

When two non-miscible solvents A and B are taken in a beaker, they form separate layers. If a solute X which is soluble in both the solvents is added, it gets distributed or partitioned between the two layers of solvents. Molecules of X pass from solvent A to B and from solvent B to A. This results subsequently a dynamic equilibrium, where the rate at which molecules of X pass from one solvent to the other is balanced (Fig. 8.1).

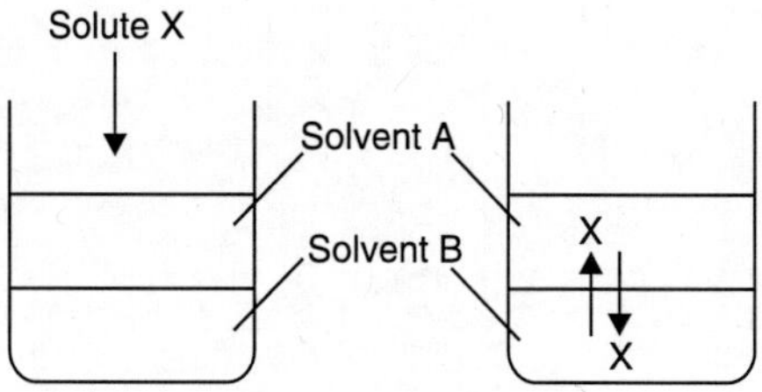

FIGURE 8.1 **Distribution of solute X between solvent A and B.**

Experimental observations show that at equilibrium.

$$\frac{\text{Concentration of X in A}}{\text{Concentration of X in B}} = \text{Constant.}$$

8.1 NERNST DISTRIBUTION LAW

At a fixed temperature, the solute to be distributed between two immiscible or slightly immiscible components or phases distributes or divides itself between two layers in a definite manner provided the distributing solute has the same molecular state in both the phases. This is known as Nernst distribution law.

For example, when iodine is added to water and carbon disulphide, it distributes itself in such a way that at equilibrium the ratio of the concentration of iodine in water to that of its concentration in carbon disulphide is constant at any given temperature.

$$\frac{[I_2]_{H_2O}}{[I_2]_{CS_2}} = \frac{C_1}{C_2} = \text{Constant}$$

where C_1 is the concentration of I_2 in H_2O

C_2 is the concentration of I_2 in CS_2

$$\frac{C_1}{C_2} = K \quad ...(i)$$

The constant 'K' is known as distribution or partition coefficient.

The distribution law also states that, "there is a mathematically constant ratio of the concentrations or the activities of the distributing solute in the two phases at fixed temperature and this ratio has been found to be independent of the amounts of the solute and the immiscible liquids."

8.2 SOLVENT EXTRACTION

One of the most important application of the distribution law is the process of solvent extraction. The extraction of an organic substance from an aqueous solution is the most important application of distribution law in the laboratory.

The solute is generally extracted from the aqueous solution by shaking the solution with a suitable organic solvent. The process is carried out by shaking the aqueous solution with a non miscible organic solvent in a separating funnel. The common solvents employed for extraction are benzene, hexane, chloroform, acetone, carbon disulphide etc.

EXPERIMENT 8.1

Aim of the Experiment

Determination of partition coefficient of iodine between water and benzene.

Apparatus Required

- Four glass stoppered reagent bottles
- Separating funnel
- Burette
- Pipette (5 and 10 ml)
- Conical flask.

Chemicals Required

- Saturated solution of iodine in benzene

- $\frac{N}{10}$ and $\frac{N}{100}$ hypo solution
- Potassium iodide
- Starch indicator.

Theory

Iodine distributes itself in two immiscible solvents as water and benzene and has the same molecular state of iodine in both of these solvents. If C_1 and C_2 are concentrations of iodine in water and benzene layers in contact with each other and forming the heterogeneous equilibrium, then the ratio of these concentrations *i.e.*, $\frac{C_1}{C_2}$ would be constant at a fixed temperature.

Mathematically

$$K = \text{Partition coefficient} = \frac{C_1}{C_2} = \frac{\text{Conc. of iodine in water layer}}{\text{Conc. of iodine in benzene layer}}.$$

Procedure

1. Take four glass stoppered bottles of about 250 ml capacity. Wash these bottles thoroughly first with a suitable reagent and then distilled water. Dry these bottles and level them as 1, 2, 3, and 4.
2. Using measuring cylinder take out 10, 20, 30 and 40 ml of iodine solution in different glass bottles.
3. Add 40, 30, 20 and 10 ml of benzene in each bottle respectively. Then add 50 ml of distilled water in each bottle.
4. Stopper these bottles and shake each bottle vigorously for about 20 minutes. Sufficient shaking of each bottle is necessary for accurate results.
5. Keep these bottles undisturbed for sometime so that two separated layers are formed distinctly. In some case droplets are found sticking to the walls of any bottle, then swirl gently the bottle to bring down these droplets.
6. Pour the contents of each bottle separately into a separating funnel and collect benzene and water layers of each bottle in separate beakers.
7. Pipette out 10 ml of organic layer from the separated layers of bottle no. 1 into a conical flask.
8. Add about 1 g of KI, 5 drops of starch solution and titrate against standard $\frac{N}{10}$ hypo solution. Disappearance of the blue colour marks the end point. Repeat this process with organic layer of other bottles in the same way.
9. Pipette out 10 ml of aqueous layer of bottle no. 1 into a titration flask. Add about 1g. of KI, 5 drops of starch solution and titrate against $\frac{N}{100}$ hypo solution. Disappearance of the blue colour marks the end point. Repeat this process with aqueous layer of other bottles in the same way.

Tabulation

No. of observations	% of iodine (V/V)	Titration with water layer			Titration with benzene layer		
		I.B.R. (ml)	F.B.R. (ml)	Vol. of $\frac{N}{100}$ hypo used (Difference V_1 ml)	I.B.R. (ml)	F.B.R. (ml)	Vol. of $\frac{N}{10}$ hypo used (Difference V_2 ml)
1	10						
2	20						
3	30						
4	40						

Calculation

$$K = \frac{\text{Concentration in } H_2O \text{ layer}}{\text{Concentration in } C_6H_6 \text{ layer}}$$

$$\text{Conc. in water layer} = \frac{\text{Normality of hypo used with water layer} \times V_1}{\text{Vol. of water layer used}} \text{ g equi/L}$$

$$\text{Conc. in benzene layer} = \frac{\text{Normality of hypo used with benzene layer} \times V_2}{\text{Vol. of benzene layer}} \text{ g equi/L}$$

values of K for each set

1. 10% solution, K =
2. 20% solution, K =
3. 30% solution, K =
4. 40% solution, K =

Conclusion

The value of K in different sets come nearly constant. Hence molecular state of iodine in water and benzene is the same.

Preparation of Reagents

1. *Sodium Thiosulphate Solution* $\left(\frac{N}{10}\right)$: Dissolve 24.82 g of $Na_2S_2O_3 . 5H_2O$ (Eq. wt. 248.21) in 1L of boiled out distilled water. Standardise it with $K_2Cr_2O_7$ to know the exact strength.
2. *Sodium Thiosulphate Solution* $\left(\frac{N}{100}\right)$: Dissolve 2.482 g of $Na_2S_2O_3 . 5H_2O$ (eq. wt. 248.21) in 1L of boiled out distilled water. Standardise it with $K_2Cr_2O_7$ to know the exact strength.
3. *Starch Indicator (1%):* Prepare a paste of 1 g of powdered starch with distilled water and add to this paste 100 ml of boiled out distilled water with constant stirring. Boil for about five minutes and then cool.

EXPERIMENT 8.2

Aim of the Experiment

Determination of the partition coefficient of benzoic acid between water and benzene at room temperature and the molecular state of benzoic acid in benzene as compared to its solution in water.

Apparatus Required

- Four reagent bottles with stoppers
- Conical flasks
- Pipette (10 ml)
- Burette
- Beaker
- Measuring cylinder.

Chemicals Required

- $\frac{N}{10}$ NaOH solution
- Benzene
- Benzoic acid
- Phenolphthalein.

Theory

If the solute associates in the solvents, the distribution law is expressed as,

$$\frac{C_1}{\sqrt[n]{C_2}} = K.$$

where n is the number of solute molecules which associate in solvent 2. Benzoic acid acts as dimer in benzene and as normal molecules in water.

Therefore, $\frac{C_1}{\sqrt{C_2}}$ should be constant.

This can be verified from experimental observations.

Procedure

1. Prepare 1% solution of benzoic acid in water.
2. Take four glass stoppered bottles of about 250 ml capacity. Wash these bottles thoroughly first with a suitable reagent and then distilled water. Dry these bottles and level them as 1, 2, 3 and 4.

3. Add using measuring cylinder into the reagent bottles as follows.

Bottle No.	*Benzene (ml)*	*1% benzoic acid solution (ml)*	*Distilled water (ml)*
1	50	50	–
2	50	40	10
3	50	30	20
4	50	20	30

4. Stopper and shake the bottles vigorously. Allow them to stand undisturbed for nearly one hour with vigorous shaking in regular intervals. Swirl the bottles gently to bring down the droplets sticking to the walls. Allow the contents of the bottle to stand till two layers separate. The upper layer is of benzene and lower layer is of water.
5. Pipette out carefully 10 ml of benzene layer from one bottle (1) and transfer it to the conical flask. Add about 20 ml of distilled water and shake. Add 1–2 drops of phenolphthalein indicator and titrate against $\frac{N}{10}$ NaOH taken in the burette until pink colour persists. Note the titre value. Repeat the process to get concordant readings. Similarly, find out the vol. of NaOH (N/10) consumed by benzene taken from remaining three bottles.
6. Close the upper end of the pipette tightly with the finger and push its nozzle into the lower water layer, in the bottle no. 1. Suck up the aqueous layer carefully upto mark (10 ml). Transfer it to the conical flask and titrate it with $\frac{N}{10}$ NaOH solution using phenolphthalein as indicator. Note down the titre value at the end point when pink colour is obtained. Repeat the titration to get concordant readings. Similarly repeat the titration of aqueous layer from the remaining three bottles.

Tabulation

	Water layer C_1			*Benzene layer C_2*		
Bottle No.	*I.B.R. (ml)*	*F.B.R. (ml)*	*Vol. of NaOH used Difference V_1 ml*	*I.B.R. (ml)*	*F.B.R. (ml)*	*Vol. of NaOH used Difference V_2 ml*
1						
2						
3						
4						

Calculation

$$K = \frac{C_1}{\sqrt{C_2}}$$

C_1 = Conc. in water layer

C_2 = Conc. in benzene layer

$$C_1 = \frac{\text{Vol. of NaOH used with water layer} \times \text{Normality of NaOH}}{\text{Volume of water layer used}} \text{ g equi/L}$$

$$C_2 = \frac{\text{Vol. of NaOH used with benzene layer} \times \text{Normality of NaOH}}{\text{Vol. of benzene layer used}} \text{ g equi/L}$$

It is observed that the values of C_1/C_2 are not constant, but those of $\frac{C_1}{\sqrt{C_2}}$ are found to be constant. Hence, the complexity of molecules of benzoic acid in benzene is 2.

Graph

Plot a graph between C_1 and $\sqrt{C_2}$. A straight line is obtained which verifies the complexity of molecules of benzoic acid in benzene as 2 (Fig. 8.2).

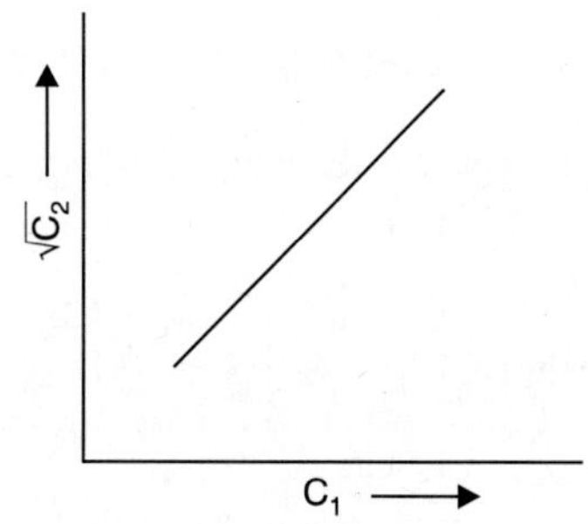

FIGURE 8.2

Conclusion

1. The value of partition coefficient between water and benzene is .... at .... °C.
2. The molecular state of benzoic acid in benzene is dimer.

Preparation of Reagents

1. *Sodium Hydroxide Solution (N/10):* Dissolve 4 g of NaOH in 1 L of distilled water.

2. *Phenolphthalein Indicator:* Dissolve 1 g of phenolphthalein in 100 ml of distilled ethyl alcohol and add 100 ml of distilled water with constant stirring. Filter if necessary.

VIVA-VOCE

1. Define the distibution law.
2. What is partition coefficient?
3. What is the application of distribution law?
4. Why do we use different concentration of the hypo solution in estimating I_2 in water and benzene layers?
5. What is meant by same molecular state of the distributing substance?
6. Explain the existence of benzoic acid in benzene and water phases.

Chapter 9

CHEMICAL KINETICS

Chemical kinetics is the branch of physical chemistry which deals with a study of the speed or velocity of the chemical reactions: Such studies also enable us to understand the mechanism by which the reactions occur. This provides very useful information on how chemical reactions occur or what is their mechanism.

9.1 RATE OF REACTION

The rate or velocity of reaction is the amount of chemical change occuring per unit time. The rate is generally defined as the decrease in concentration of reactant or increase in concentration of the products per unit time. Mathematically, the rate of reaction can be expressed as,

$$\text{Rate of reaction} = \frac{\text{Amount of reactant consumed}}{\text{Time interval}} = \frac{\text{Amount of product obtained}}{\text{Time interval}}$$

For a reaction,

$$A \rightarrow B$$

The rate of reaction is given by,

$$r = -\frac{d[A]}{dt} = +\frac{d[B]}{dt}$$

where, [A] and [B] indicates concentration of A and B respectively. It must be noted that the rate of reaction is invariably a positive quantity. The negative sign merely indicates that concentration of the reactant is falling with increase in time while the positive sign indicates the increase in concentration of the product with increase in time.

Rate has the units of concentration divided by time *i.e.* mol L^{-1} sec^{-1}.

9.2 RATE CONSTANT

Consider the following reaction

$$A \rightarrow B$$

$$\text{Rate of reaction, } r = \frac{dx}{dt} = -\frac{d[A]}{dt} = k[A]$$

where [A] represents the concentration of reactant A and k is the proportionality constant called velocity constant, rate constant or specific reaction rate at given temperature. It can be defined as fraction of A that reacts per unit time

If $[A] = 1$, then $r = k$

Thus at a given temperature, and when the concentration of the reactant is unity the rate constant of a reaction involving a single reactant is equal to the rate of the reaction.

9.3 ORDER OF REACTION

The order of reaction is defined as the total number of reactants whose concentrations change during the course of a reaction. It is the sum of exponents of the molar concentration of the reactants in the rate equation of the reaction.

e.g.,

$$mA + nB \longrightarrow \text{Products}$$

$$\text{Rate, } \frac{dx}{dt} = k[A]^m\,[B]^n$$

$$\text{order of reaction} = m + n.$$

Thus the order of reaction can be defined as the sum of the power of the concentration terms.

$m + n = 1$ (First order reaction)

$m + n = 2$ (Second order reaction)

$m + n = 3$ (Third order reaction)

Zero order reaction is one whose rate is independent of concentration.

Order of a reaction can not be obtained from a simple balanced equation. It can be known by experimental means. It may be a whole number, fraction or zero.

9.4 MOLECULARITY OF REACTION

Molecularity of a reaction is defined as the minimum number of atoms or molecules or ions of the reactants taking part in the rate determining step.

e.g.,

$$N_2O_5 \xrightarrow{\Delta} N_2O_4 + \frac{1}{2}O_2 \qquad \text{(Molecularity=1)}$$

$$2HI \longrightarrow H_2 + I_2 \qquad \text{(Molecularity = 2)}$$

$$2NO + O_2 \longrightarrow 2NO_2 \qquad \text{(Molecularity = 3)}$$

Molecularity of a reaction is always a whole number and is never a fraction or zero.

Kinetics of First Order Reaction

$$A \rightarrow \text{Product}$$

$$k = \frac{2.303}{t} \log \frac{a}{a-x}$$

Kinetics of Second Order Reaction

1. The rate is proportional to the square of the concentration of the same reactant.

   Consider the reaction, $2A \rightarrow$ Products

$$k = \frac{1}{t} \times \frac{x}{a(a-x)}$$

2. The rate is proportional to the product of the concentration of two reactants.

   Consider the reaction: $A + B \rightarrow$ Products

$$k = \frac{2.303}{t(a-b)} \log \frac{b(a-x)}{a(b-x)}$$

Kinetics of Third Order Reaction

$$3A \rightarrow \text{Product}$$

$$k = \frac{1}{t} \times x\,(2a - x)/2a^2(a-x)^2$$

EXPERIMENT 9.1

Aim of the Experiment

To find out velocity constant for the inversion of cane sugar in acidic medium and to show that inversion follows the first order kinetics.

Apparatus Required

- Polarimeter
- Beaker
- Thermostat
- Stop watch
- Conical flask.

Chemicals Required

- Pure cane sugar crystals
- 2M HCl

Theory

The inversion of cane sugar takes place in presence of H^+ ions as follows:

$$\underset{\substack{\text{Sucrose}\\\text{dextrorotatory}}}{C_{12}H_{22}O_{11}} + H_2O \longrightarrow \underset{\text{Glucose}}{C_6H_{12}O_6} + \underset{\substack{\text{Fructose}\\\text{laevorotatory}}}{C_6H_{12}O_6}$$

Above equation shows that dextrorotatory sucrose is hydrolysed in presence of acid into laevorotatory glucose and fructose. Kinetically the reaction is of first order, since water is taken in excess and hence the concentration of H_2O remains practically constant. The rate of the reaction depends entirely upon the concentration of cane sugar. The reaction is pseudo unimolecular.

The kinetics of this reaction can be studied using polarimeter by measuring the angle of rotation of plane of polarised light through the reaction mixture which changes as reaction proceeds. The angle of rotation of reaction mixture gradually reaches to zero from positive value and then goes to negative.

Suppose,

r_0 = Angle of rotation at zero time

r_t = Angle of rotation at time 't'

r_∞ = Angle of rotation at infinity.

The rate expression for first order kinetics is given by

$$k = \frac{2.303}{t} \log \frac{a}{a-x}$$

where, K = Velocity constant

a = Initial concentration of cane sugar

x = Amount of cane sugar hydrolysed in time 't'

For the inversion of cane sugar, 'a' will be proportional to total change in the angle of rotation.

$$a = r_0 - r_\infty$$

Similarly, concentration of sugar left at any time 't'

$$a - x = r_t - r_\infty$$

Now, the equation can be written as,

$$k = \frac{2.303}{t} \log \frac{r_0 - r_\infty}{r_t - r_\infty}.$$

Procedure

Weigh 20 g of pure cane sugar crystals in clean, dried weighing bottle and transfer into 100 ml of volumetric flask. Dissolve in distilled water and make up the volume up to mark and shake well. Transfer 25 ml of sugar solution into a dry conical flask. Take 40 ml of 2M HCl in another conical flask. Place the two in thermostat maintained at 30°C.

Start polarimeter and determine the zero reading with distilled water at room temperature. Remove water from the polarimeter tube and dry it. Transfer the 25 ml sugar solution which was placed in thermostat into this tube and then quickly add 25 ml of 2M HCl, immediately record the value of angle of rotation (r_0). Start stop watch and note down the reading after time intervals of 5, 10, 15, 20, 30, 40 and 60 minutes. Then, warm the mixture to 70°C for an hour to complete the reaction, cool and note the reading to get value of r_∞.

Tabulation

Temperature of the experiment = t°C

No. of observations	*Time in minutes*	*Rotation* r_t	*Angle of rotation* $\log(r_t - r_\infty)$	*Velocity constant (K) minutes*$^{-1}$
1	0	r_0		
2	5			
3	10			
4	15			
5	20			
6	30			
7	40			
8	60			
9	∞	r_∞		

Observation

As the value of k is the same at different time intervals, this reaction is therefore of first order.

Graph

Plot a graph between log ($r_t - r_\infty$) vs time 't'.

A straight line shows the first order kinetics of the reactions.

Conclusion

1. Velocity constant for the inversion of cane sugar in acidic medium at t°C = ... minutes^{-1}
2. A straight line of the graph shows that the reaction follows first order kinetics.

Preparation of Reagents

1. *Hydrochloric Acid (2M):* Dilute 500 ml concentrated HCl (12 M) to 3L with distilled water.

EXPERIMENT 9.2

Aim of the Experiment

To determine the value of rate constant (K) for the hydrolysis of ethyl acetate catalysed by hydrochloric acid.

Apparatus Required

- Burette
- Pipette (5 ml, 10 ml)
- Conical flask (six 50 ml and one 250 ml)
- Stop watch
- Thermostat.

Chemicals Required

- Ethyl acetate
- $\frac{N}{2}$ HCl solution
- $\frac{N}{10}$ NaOH solution
- Phenolphthalein indicator.

Theory

The hydrolysis of ester (ethyl acetate) in presence of dilute acid is an example of first order reaction, as the concentration of water is in large excess and does not change significantly during the course of the reaction. This is therefore pseudo-unimolecular reaction.

$$\underset{\text{Ethyl acetate}}{CH_3COOC_2H_5} + H_2O \xrightarrow{H^+} \underset{\text{Acetic acid}}{CH_3COOH} + \underset{\text{Ethyl alcohol}}{C_2H_5OH}$$

The velocity constant is calculated from the equation of first order reaction.

$$k = \frac{2.303}{t} \log \frac{a}{a-x}$$

where, a = Initial concentration

$(a - x)$ = Concentration after time 't'

The reaction is acid catalysed in presence of HCl.

Since during hydrolysis free acetic acid is produced, the reaction is investigated by titrating so formed acetic acid against standard alkali at different time 't'.

If, V_0 = Initial concentration when $t = 0$

V_t = Titration after different timings

V_∞ = Final titration *i.e.* when hydrolysis is complete ($t = \infty$)

Then,
$$a \propto (V_\infty - V_0)$$
$$(a - x) \propto (V_\infty - V_t)$$
$$k = \frac{2.303}{t} \log \frac{(V_\infty - V_0)}{(V_\infty - V_t)}.$$

Procedure

Take 100 ml of $\frac{N}{2}$ HCl in a 250 ml conical flask, stopper it and keep it in a thermostat or water bath at room temperature. Take 20 ml of ethyl acetate in 50 ml conical flask, stopper and keep the flask in the same water bath or thermostat. Allow to stand the solutions for 20 minutes to attain bath temperature. In the mean time fit the burette properly and fill it with 0.1 N NaOH solution. Also add 15 g of pounded ice in each of the six 50 ml conical flasks.

Pipette out 5 ml of the ester from the conical flask and add it to the flask containing 100 ml of 0.5 N HCl.

Shake the contents, pipette out 10 ml of the reaction mixture and transfer it at once to first conical flask containing pounded ice. Titrate this against 0.1 N NaOH solution taken in the burette after adding a drop of phenolphthalein indicator. Appearance of pale pink colour marks the end point. Note down the titre value as V_0.

After 10 minutes of the initial withdrawal, again pipette out 10 ml of the reaction mixture and add to the second conical flask containing ice, shake well and titrate against $\frac{N}{10}$ NaOH as described above. Note down the titre value V_t after 10 minutes.

Repeat the same procedure after every 10 minutes so as to get a total of six readings.

Finally remove the flask from the thermostat or water bath and place it in a separate water bath at a temperature of 50–60°C for about one hour. Pipette out 10 ml of the reaction mixture and titrate against $\frac{N}{10}$ NaOH with phenolphthalein as indicator. Note down the titre value as V_∞.

Tabulation

Temperature of the water bath = t°C

No. of observations	*Time in minutes*	*I.B.R. (ml)*	*F.B.R. (ml)*	*Vol. of NaOH solution used Difference in ml*	
1	0			V_0	$V_\infty - V_0$
2	10			V_{10}	$V_\infty - V_{10}$
3	20			V_{20}	$V_\infty - V_{20}$
4	30			V_{30}	$V_\infty - V_{30}$
5	40			V_{40}	$V_\infty - V_{40}$
6	50			V_{50}	$V_\infty - V_{50}$
7	∞			V_∞	

Calculation

$$k = \frac{2.303}{t} \log . \frac{a}{a-x}$$

$$= \frac{2.303}{t} \log \frac{V_\infty - V_0}{V_\infty - V_t} \text{ minute}^{-1}$$

Calculate the value of rate constant (k) at different time intervals.

Graph

The rate equation for the first order reactions

$$t = \frac{2.303}{K} \log \frac{a}{a-x}$$

$$= \frac{2.303}{K} \log a - \frac{2.303}{K} \log (a-x)$$

$$= -\frac{2.303}{K} \log (a-x)$$

It indicates that a plot of t against log $(a - x)$ or log $(V_\infty - V_t)$ will be a straight line whose slope is given by $-\frac{2.303}{K}$.

Thus,
$$\frac{-2.303}{K} = \tan\theta = -\frac{OA}{OB}$$

or
$$k = 2.303 \times \frac{OB}{OA}$$

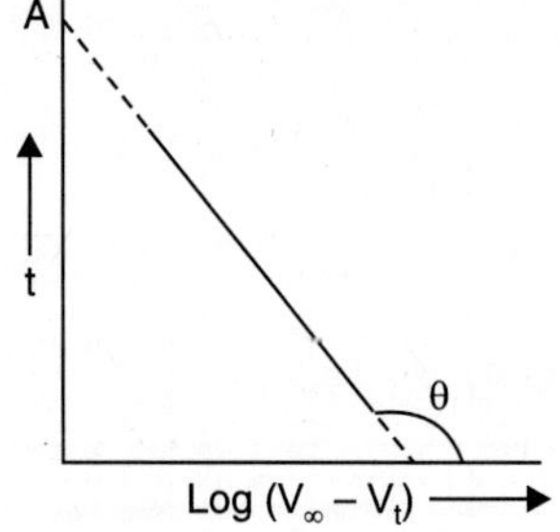

FIGURE 9.1 **Plot of t $(V_\infty - V_t)$.**

Conclusion

The hydrolysis of ester (ethyl acetate) catalysed by dilute acid is a first order reaction because the value of k is constant and the graph is a straight line.

Preparation of Reagents

1. *Hydrochloric Acid Solution* $\left(\frac{N}{2}\right)$: Dilute 45 ml of conc. HCl to 1 L with distilled water and standardise with $\frac{N}{2}$ sodium carbonate solution using methyl orange indicator.
2. *Sodium Hydroxide Solution* $\left(\frac{N}{10}\right)$: Dissolve 4 g of NaOH in 1 L of distilled water.
3. *Phenolphthalein Indicator:* Dissolve 1 g of phenolphthalein in 100 ml of distilled ethyl alcohol and add 100 ml of distilled water with constant stirring. Filter if necessary.

VIVA-VOCE

1. Define the term order and molecularity of a reaction.

2. What is the significance of chemical kinetics?

3. What is velocity of reaction?

4. Give examples of first and second order reactions.

5. Give the units of rate constant for first and second order reactions.

6. Why are third or higher order reactions very rare?

Ans. Third or higher order reactions are very rare because in such reactions more reacting molecules should meet simultaneously which is improbable.

Chapter 10 PREPARATIONS

EXPERIMENT 10.1

Aim of the Experiment

Preparation of phenolphthalein.

Apparatus Required

- Round-bottomed flask (250 ml)
- Beaker (2L)
- Buchner funnel
- Oil bath
- Water bath.

Chemicals Required

- Phenol
- Phthalic anhydride
- Conc. H_2SO_4
- Dil. NaOH
- Absolute alcohol
- Dil. HCl
- Dil. CH_3COOH.

Theory

Phenol condenses with phthalic anhydride in the presence of conc. H_2SO_4 or anhydrous $ZnCl_2$ to yield the colourless phenolphthalein as the main product. When dilute caustic alkali is added to an alcoholic solution of phenolphthalein, an intense red colouration is appeared. The alkali opens the lactone ring in phenolphthalein and forms a salt at one phenolic group. The reaction may be represented in steps, with the formation of a hypothetical unstable intermediate that changes to a coloured ion. The colour is probably due to resonance which places the negative charge on either of the two equivalent oxygen atoms. With excess of conc. caustic alkali, the first red colour disappears; this is due to the production of the carbinol and attendant salt formation, rendering resonance impossible. The various reactions may be represented as follows:

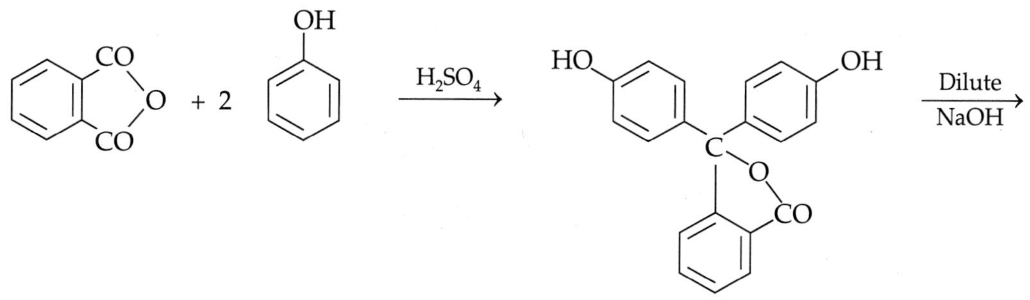

Phenolphthalein (colourless)

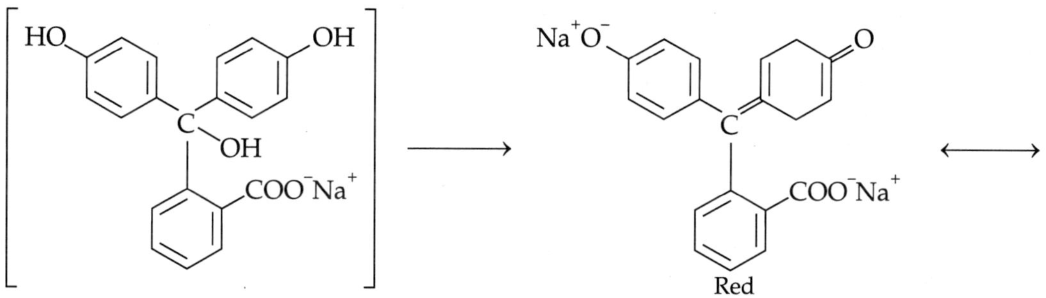

Red

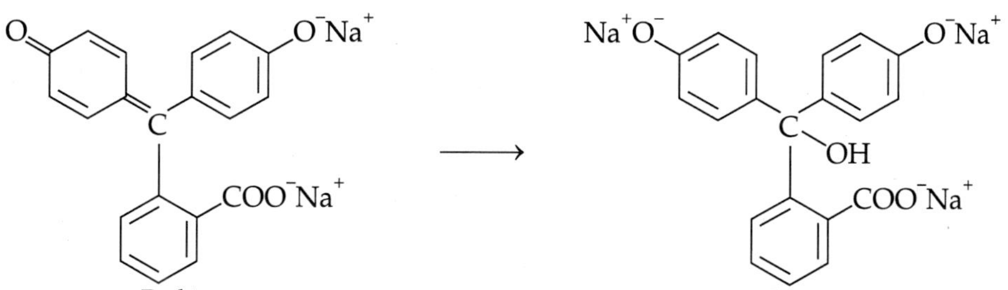

Red Colourless

Procedure

To a mixture of pure phenol and 25 g of phthalic anhydride contained in a 250 ml round bottomed flask, add 20 g (11 ml) of concentrated sulphuric acid. Heat the flask in an oil bath at 115–120°C for 9 hours. Then pour the reaction mixture whilst still hot into 1 litre of hot water contained in a 2 litre beaker, and boil until the odour of phenol has disappeared; add water to replace that lost by evaporation. When cold, filter the yellow, granular precipitate at the pump and wash it with water. Dissolve the solid in dilute sodium hydroxide solution, filter from the undissolved residue (the by-products of the reaction).

Acidify the filtrate with dilute acetic acid and a few drops of dilute hydrochloric acid and allow to stand overnight. The crude phenolphthalein separates as a pale yellow, sandy powder; filter and dry. Purify the crude product by dissolving it in six times its weight of absolute alcohol, and decolourising carbon and reflux on a water bath for one hour. Filter the hot solution through a preheated buchner funnel, wash the residue with 2 parts by weight of boiling absolute alcohol and concentrate the combined filtrate and washings to two-thirds of its bulk on a water bath. Dilute the cooled solution with eight times the weight of cold water (it will become turbid), stir the mixture well and after standing for a few seconds, filter through a wet filter to remove the resinous oil which separates. Heat the filtrate on a water bath to evaporate most of the alcohol; the turbidity disappears and the phenolphthalein separates out in the form of a white powder. Filter this off and dry.

Conclusion

Yield = 18 g

The melting point of phenolphthalein was found to be 256–258°C (Pure).

Preparation of Reagents

1. *Dilute Acetic Acid (5N):* Dilute 287 ml. of glacial acetic acid to 1L of distilled water.
2. *Dilute Hydrochloric Acid (5N):* Dilute 430 ml. of concentrated HCl to 1L with distilled water.

EXPERIMENT 10.2

Aim of the Experiment

Preparation of Methyl orange (Azodye).

Apparatus Required

- Ice bath
- Beaker
- Conical flask (250 ml)
- Test tube.

Chemicals Required

- N, N-dimethylaniline
- Sulphanilic acid
- Sodium nitrite
- Hydrochloric acid
- Acetic acid
- Sodium hydroxide solution (20%)
- Sodium chloride
- Sodium carbonate.

Theory

Dyes may be defined as the chemical substances which when applied to a fibre or a surface impart a relatively permanent colour to it. All the dyes which have one/more azo groups (—N=N—) in their structure are called azo dyes.

Preparation of an azodye involves two stages.

1. *Diazotisation:* In the diazotisation step a primary aromatic amine reacts with nitrous acid (HONO) at a temperature of 0-5°C to form a diazonium salt. The reaction is known as diazotisation. The nitric acid is generated in situ by sodium nitrite and a mineral acid like HCl. In the preparation of methyl orange, sulphanilic acid is diazotised and coupled to N, N-dimethylaniline. The reaction of diazotisation may be represented as follows:

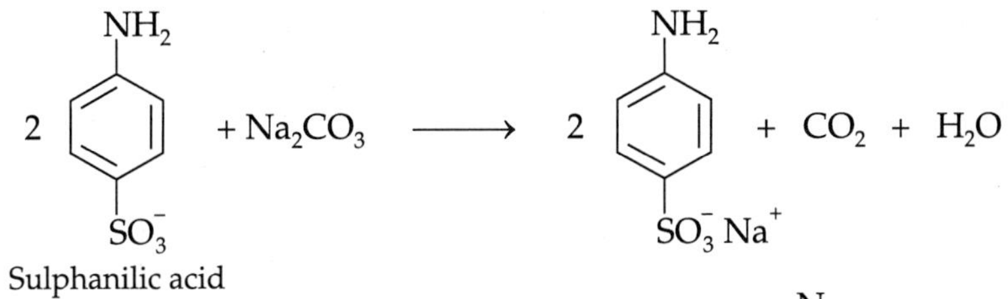

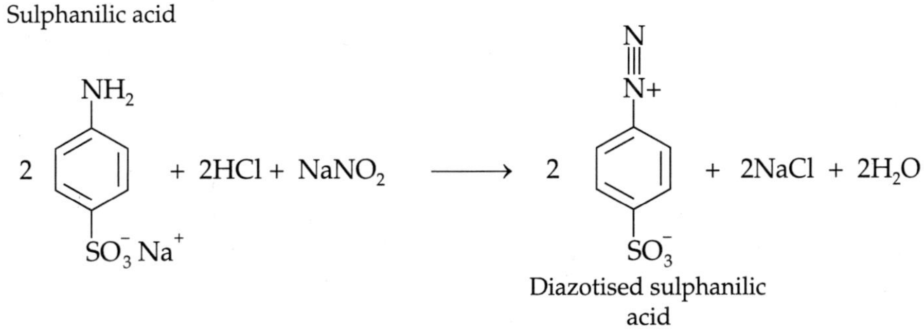

2. *Coupling:* Diazonium salts are soluble and reactive compounds. On condensation with aromatic amines in acidic medium, they produce highly coloured compounds (azo dye). The reaction being known as coupling reaction. As diazonium salts are very unstable so the reaction is carried out at low temperature (below 5°C). These are not isolated from the solution and the coupling agent is added directly in diazonium salt solution.

Procedure

Diazotisation

1. Dissolve 7 g of finely powdered sulphanilic acid and 2 g of sodium carbonate in 50 ml of water. Warm the mixture gently to obtain a clear solution.
2. Cool the flask under tap water and to it add a solution of 2.2 g of sodium nitrite in 10 ml of water.
3. Cool the flask in ice bath until the temperature falls to 5°C. Now add very slowly, drop by drop a solution of 8 ml of conc. HCl in 15 ml of water to the flask. Stir continuously. Do not allow the temperature to rise above 10°C.
4. Allow the solution to stand in ice water for 15 minutes to ensure complete diazotisation. A solid material (4-diazobenzene sulphuric acid), starts separating slowly.

Coupling

5. Dissolve 4 ml of dimethylaniline in 3 ml of glacial acetic acid diluted with 10 ml of water and cool the solution in ice water.
6. Add the solution of dimethylaniline acetate, obtained in step 5, to the flask containing diazotised sulphanilic acid prepared in step 4.
7. Allow the reaction mixture to stand for 10 minutes when crystals of methyl orange start separating.
8. Add approximately 20 ml of 20% solution of sodium hydroxide to convert the dye into its yellow sodium salt. Add 5 g of pure NaCl and warm the solution to 60°C.
9. Shake and cool the mixture in an ice bath, filter the dye under a very gentle pressure.
10. Wash with cold water, dry and recrystallise from boiling water, weigh and report the yield.

Conclusion

Yield = ... g.

Preparation of Reagents

1. *Sodium Hydroxide Solution (20%):* Dissolve 20 g of NaOH in 100 ml of distilled water.

EXPERIMENT 10.3

Aim of the Experiment

Preparation of acetyl salicylic acid (Aspirin).

Apparatus Required

- Conical flask
- Beaker (100 ml)
- Buchner funnel.

Chemicals Required

- Salicylic acid
- Acetic anhydride
- Conc. H_2SO_4.

Theory

Salicylic acid upon acetylation yields acetyl salicylic acid or aspirin. Acetylation proceeds readily with acetic anhydride in presence of a little conc. H_2SO_4 as catalyst.

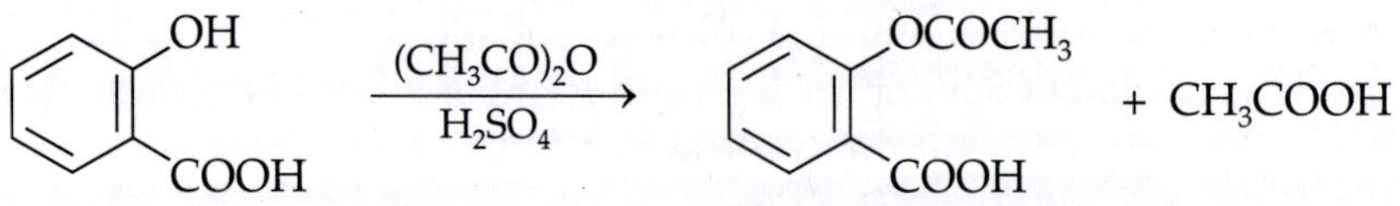

Solicylic acid Acetyl salicylic acid

Procedure

Take 5 g of salicylic acid, 7 ml of acetic anhydride in a 100 ml conical flask. To it add 2–3 drops of conc. H_2SO_4, shake and then heat the flask on a water bath keeping temperature between 60–70°C for 20 minutes. Cool the mixture in an ice bath with stirring, add about 20 ml of ice cold water to decompose the excess of acetic anhydride. Pour the contents of the conical flask into a 250 ml beaker containing 50 g of crushed ice, stir until white solid appears. Scratch the side of the flask with a glass rod or stainless steel spatula if white solid does not appear, filter through buchner funnel, dry by pressing into the folds of filter paper. Recrystalise the crude acetyl salicylic acid from a mixture of equal volumes of acetic acid and water.

Conclusion

Yield = 5 g

M.P = 136 – 137°C

Acetyl salicylic acid decomposes when heated and does not possess a true clearly defined melting point.

10.1 POLYMERS

The polymers are emerging as very important engineering materials. The word polymer is derived from the Greek 'Poly' or many and 'meros' or parts. Thus polymer means composed of many parts or many units. Polymer is defined as a macromolecule of high molecular weight consisting of repeating units held together by covalent bond. The starting material from which the molecule is formed is known as the monomer. The repeating unit in the polymer is called a mer. A polymer is a chain of repeating mers like beads in a necklace. In other words, a polymer molecule can be defined as a very long molecule in which a simple chemical unit repeats itself a large number of times.

Organic polymers can be divided into two broad divisions of thermoplastics and thermosetting plastics. Thermoplastics soften under the action of heat and harden on cooling and can be remoulded further. A thermosetting material is one which takes up its permanent shape when heat and pressure are applied during the manufacturing process. It can not be softened when reheated and hence the thermosetting resins can not be reformed and remoulded. Thermosetting resins include phenolics (bakelite), amino resins, polyester etc. Thermoplastics include polystyrenes, polyamides, polyethylene etc.

Polymerisation: The process by which polymers are synthesised is called polymerisation. Polymerisation is carried out by following means.

1. *Addition Polymerisation:* In this type of polymerisation, the product so formed is an exact multiple of the original monomeric molecule. Such a polymerisation process involving addition of monomer units to the growing chain is known as chain growth polymerisation.

e.g.,

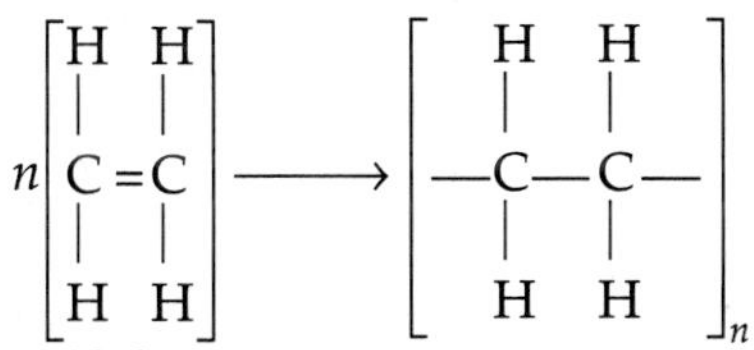

Ethylene Polyethylene

2. *Condensation polymerisation:* In this type of polymerisation the polymer is formed through reaction of functional groups. By products like H_2O, HCl are eliminated during such process. The process takes place in stepwise manner and hence known as step growth polymerisation.

e.g.,

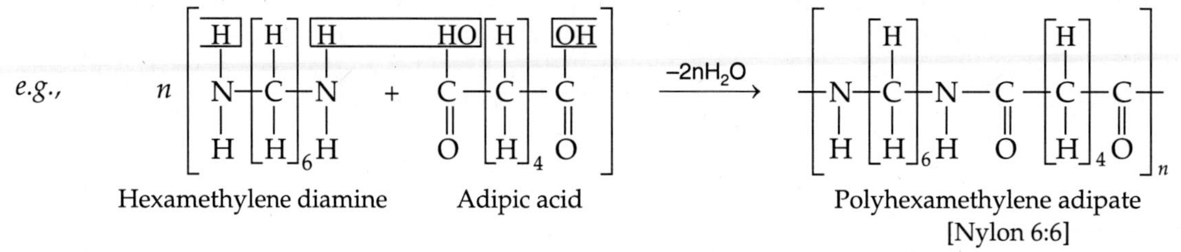

3. *Copolymerisation:* This is the process in which a mixture of two or more different monomers gets polymerised to yield a product. The product is entirely a new product and is called copolymer.

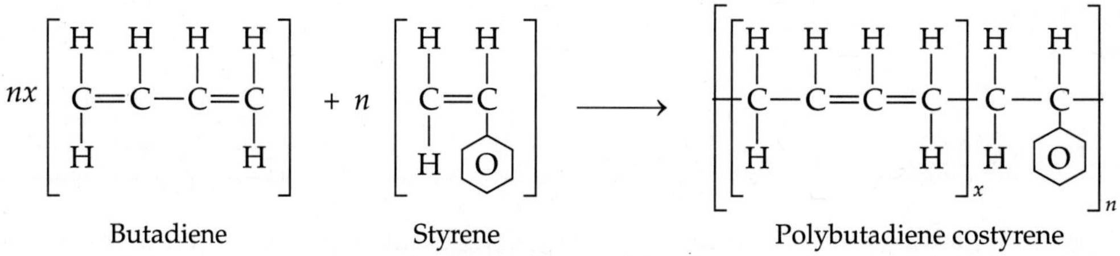

EXPERIMENT 10.4

Aim of the Experiment

Preparation of phenol-formaldehyde resin (Bakelite).

Apparatus Required

- Beaker
- Funnel
- Glass rod.

Chemicals Required

- Glacial acetic acid, 40% formaldehyde solution, phenol, conc. HCl.

Theory

Phenolic resins are condensation polymerisation products of phenolic derivatives with aldehydes. Most important member of this class is bakelite or phenol-formaldehyde resin. It is prepared by condensing phenol with formaldehyde in presence of acidic/alkaline catalyst.

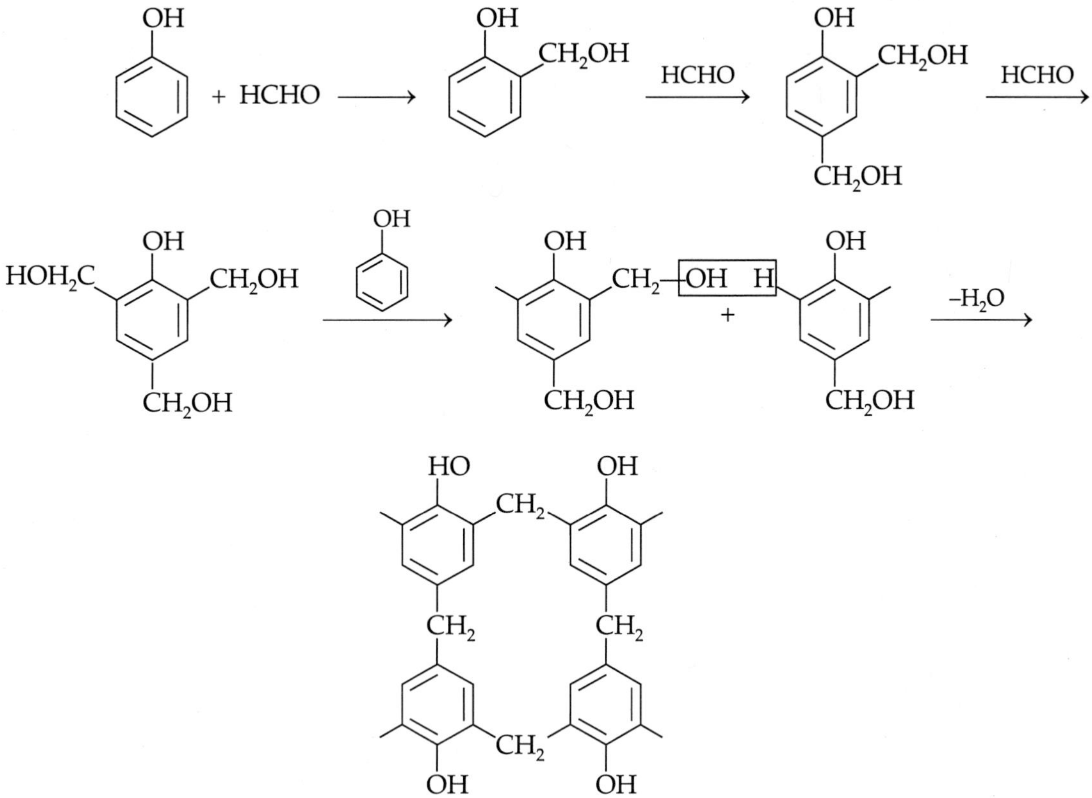

Bakelite

Procedure

Take 5 ml of glacial acetic acid and 2.5 ml of 40% formaldehyde solution in a 500 ml beaker and add 2 g of phenol. Wrap a cloth or towel loosely round the beaker. Pass dry HCl gas (or a few ml. of conc. HCl) into the mixture carefully. Within 5 minutes, a large mass of pink plastic is formed. Heat slightly if conc. HCl is being used. The residue obtained is washed several times with distilled water and filtered product is dried and weighed.

Conclusion

Yield of the product is ...... g.

Preparation of Reagents

1. *Formaldehyde Solution (40%):* Dilute 40 ml of formaldehyde to 100 ml with distilled water.

VIVA-VOCE

1. What is carbolic acid?
2. Why it is not advisable to take the melting point of dye?

   Ans. The dye decomposes before it melts. Hence it is not advisable to take the melting point of dye.

3. Define Dye.
4. What is acetylation?
5. What are acetylating agents?

   Ans. Acetyl chloride and acetic anhydride.
6. What are polymers?
7. What are the uses of phenol-formaldehyde resins?

   Ans. They find use in telephone parts, radio and TV cabinets, automobile parts, electric switches and plugs, varnishes, paints, cooker handle etc.
8. What are the types of polymerisation?
9. What is the difference between thermosetting and thermoplastic polymer?
10. What are the applications of Bakelite?

Chapter 11 INSTRUMENTATION

11.1 pH MEASUREMENT

pH a measure of hydrogen ion activity is used to express the intensity of acidic or alkaline condition of a solution, because most of the chemical and biochemical processes are profoundly affected by the acidity or alkalinity of the medium. Therefore pH is an important factor in any kind of analysis.

Sorensen suggested the use of pH scale. On this scale pH of any solution is defined as

$$pH = -\log_{10} aH^+$$

where aH^+ is the activity of the hydrogen ions. For dilute solutions activity can be replaced by concentration.

$$pH = -\log_{10} [H^+]$$

pH scale is logarithmic. The pH of a solution is defined as the negative logarithm of the hydrogen ion concentration. The pH of a solution can be accurately measured with the help of a pH meter. However the approximate value of pH can also be determined using a multiple range indicator solution.

11.2 DETERMINATION OF pH

For determining the pH value an electrode reversible to H^+ ions is required. Various electrodes like hydrogen electrode, quinhydrone electrode etc., can be employed. But glass electrode is the best amongst all these electrodes. The determination of H^+ concentration depends on the fact that when a thin glass membrane is in contact with solutions of different hydrogen ion concentrations on its two sides, a potential difference is developed across the glass membrane. The magnitude of the potential difference depends on the difference in the concentration of the hydrogen ions in two solutions. The reference electrode employed is usually the calomel electrode. Instruments are available in which the galvanometer readings are directly scaled to read pH values. These instruments are called pH meters. The cell may be represented as follows.

Pt. 0.1 N HCl/glass/Experimental solution/KCl saturated solution Hg_2Cl_2(s)solution/Hg

Glass electrodes act as the negative electrode and the calomel electrode act as the positive electrode.

$$pH = \frac{E_{cell} - 0.2420 + E^\circ_G}{0.0591}.$$

EXPERIMENT 11.1

Aim of the Experiment

Preparation of buffer solution and determination of pH of a buffer solution.

Apparatus Required

- pH meter with glass electrode and standard
- Calomel electrode
- Beaker (100 ml).

Chemicals Required

- Buffer solution (pH 4 and 9)
- Ammonium chloride
- Ammonium hydroxide
- Ammonium acetate
- Glacial acetic acid
- $\frac{M}{5}$ acetic acid
- $\frac{M}{5}$ sodium acetate
- $\frac{N}{5}$ hydrochloric acid
- N/5 potassium chloride.

Theory

The modern pH meter is an electronic digital voltmeter scaled to read pH directly. All pH meters have provision for standardising the glass electrode in a buffer solution of known pH. This is necessary because different electrodes have different asymmetry potentials. Once the adjustment has been made so that the meter registers correctly the known pH of the buffer solution, the instrument gives the pH of other solutions without calculation.

Procedure

1. *Preparation of Buffer Solutions:* Prepare the buffer solutions having pH = 4 and pH = 9 (buffer tablets BDH and E-Merck are available). These can also be prepared in the laboratory. 0.05 M solution of potassium hydrogen phthalate gives a buffer solution of pH 4.008 at 25°C, 0.05 M solution of borax gives a buffer solution of pH 9.18 at 25°C.

   (*i*) Ammonium chloride-Ammonium hydroxide buffer solution (pH = 10):

   Dissolve 70 g NH_4Cl and add 568 ml of Conc. ammonia solution having specific gravity 0.88 – 0.90. Stir and make up the volume to 1 L with distilled water.

   (*ii*) Ammonium acetate buffer solution (pH = 5.0):

   Dissolve 250 g of ammonium acetate in 150 ml of distilled water. Add 700 ml of glacial acetic acid.

   (*iii*) 41.0 ml M/5 acetic acid + 9.0 ml M/5 sodium acetate (pH = 4.0)

   (*iv*) 5.3 ml of N/5 HCl + 25 ml N/5 KCl. Dilute the mixture to 100 ml with distilled water (pH = 2.0).

2. *Determination of pH:*

(*i*) Switch on the instrument and allow the instrument to warm up for 10–15 minutes.

(*ii*) Wash the calomel and glass electrodes (or combined electrode) with distilled water.

(*iii*) Take the buffer solution (pH = 4) in a clean glass beaker. Lower the electrodes so that they are immersed in the solution to a depth of about one inch.

(*iv*) Measure the temperature of the solution by a thermometer and set temperature compensate control to this value.

(*v*) Insert the electrode assembly into the same beaker and set the selector switch (if available) of the instrument to read pH range (0–7).

(*vi*) Adjust the 'set buffer' control until the meter reading agrees with the known pH of the buffer solution (pH = 4).

(*vii*) Put back the selector to zero position. Remove the electrode assembly, rinse in distilled water, dry with tissue paper and place into a small beaker containing second buffer solution (pH = 9). Same procedure is to be adopted except now the selector switch is put to a pH range of 7–14.

(*viii*) Remove the electrode assembly, rinse in distilled water followed by drying with tissue paper. Place in the first buffer solution and confirm that the correct pH reading is shown on the meter. If not, repeat the calibration procedure.

(*ix*) If the calibration is satisfactory, rinse the electordes etc., with distilled water and introduce into the test solution (buffer solutions) contained in a small beaker. Read off the pH of the solution.

(*x*) Remove the electrodes etc. Rinse in distilled water and leave standing in distilled water.

Conclusion

The pH of the buffer solutions prepared is recorded as

(*i*) ....

(*ii*) ....

(*iii*) ....

(*iv*) ....

Preparation of Reagents

1. *Acetic Acid (M/5):* Dilute 11.5 ml of 17.4 N and 99.5% glacial acetic acid to 1 L with distilled water in a volumetric flask.
2. *Hydrochloric Acid (N/5):* Dilute 18 ml of conc. hdyrochloric acid to 1 L with distilled water and standardise with Na_2CO_3 (N/5).
3. *Potassium Chloride (N/5):* Dissolve 14.9 g of KCl in distilled water and dilute to one litre.
4. *Sodium Acetate (M/5):* Dissolve 16.4 g of sodium acetate in distilled water and dilute to one litre.

EXPERIMENT 11.2

Aim of the Experiment

Determination of the strength of hydrochloric acid solution pH metrically against sodium hydroxide solution of known strength.

Apparatus Required

- pH meter with glass electrode and reference electrode
- Beaker
- Burette.

Chemicals Required

- HCl solution
- N/10 NaOH solution.

Theory

When an alkali is added to an acid solution, the pH of the solution increases slowly. But at the equivalence point, the rate of change of the solution is very rapid. A plot is drawn between volume of the alkali added and the pH of the solution. The sharp break in the curve gives the equivalence point, from which the strength can be calculated using normality equation.

Description of the Instrument: The modern pH meter is an electronic digital voltmeter, scaled to read pH directly. All pH meters have provision for standardising the glass electrode in a buffer solution of known pH. This is necessary because different electrodes have different asymmetry potentials. Once the adjustment has been made; so that the meter registers correctly the known pH of the buffer solution, the instrument gives the pH of other solution without any calculation.

Procedure

1. Calibrate the pH meter with the glass electrode in the buffer solution of known pH.
2. Standardise the pH meter with acidic buffer (pH = 4) as above. Wash the electrode with distilled water and dry.

3. Take 50 ml of HCl solution in a beaker and dip the electrode completely into it.
4. Note the pH of the pure acid solution.
5. Add to this solution 1 ml of N/10 NaOH solution from the burette, stir well and note down the pH of the solution.
6. Continue adding NaOH solution from the burette and note down the pH after each addition.
7. Near the equivalence point the change in pH is much more rapid than in any other region.

Tabulation

Volume of HCl taken = 50 ml.

Volume of alkali added (ml)										
pH										

Calculation

Plot a curve with pH values as ordinate and the volume of alkali added as abscissa. The sharp break in the curve corresponds to the equivalence point.

Let the volume of alkali at equivalence point = x ml

$$\underset{\text{Acid}}{N_1 V_1} = \underset{\text{Alkali}}{N_2 V_2}$$

$$N_1 \times 50 = \frac{N}{10} \times x$$

$$N_1 = \frac{N}{10} \times \frac{x}{50}$$

$$\text{Strength of HCl solution} = \frac{36.5 \times x}{10 \times 50} \text{ g/L.}$$

Conclusion

The strength of the given acid solution is $\frac{36.5 \times x}{10 \times 50}$ g/L.

Preparation of Reagents

1. *Sodium Hydroxide Solution (N/10):* Dissolve 4 g of NaOH in 1L of distilled water.

11.3 CONDUCTOMETRIC MEASUREMENT

Conductometry deals with the measurement of electrical conductance of solutions containing electrolytes (acids, bases and salts in water).

Measurement of Conductivity: The electrolytic conductance of an electrolyte is measured in "conductivity cells" which are specially designed vessels of suitable size and shape in which two plates of an inert metal (*e.g.* Pt or Au) are provided to serve for the flow in and flow out of electrons. The electrolyte whose conductivity is to be measured is taken either in fused state or in aqueous solution. In order to measure the conductivity of the solution the solution is placed in a conductivity cell of which the cell constant has been determined by calibration with a solution of accurately known conductivity. A standard KCl solution is usually employed for this purpose.

EXPERIMENT 11.3

Aim of the Experiment

To find out the percentage purity of $AgNO_3$ solution conductometrically, against N/10 KCl solution.

Apparatus Required

- Conductometer
- Conductivity cell.

Chemicals Required

- Conductivity water
- N/10 KCl solution
- Given $AgNO_3$ solution.

Theory

Conductometric titration is the volumetric analysis based upon the measurement of conductance during the course of titration.

The titration of $AgNO_3$ vs KCl *i.e.*, precipitation reactions can be carried out with the help of conductometric measurement methods.

$$Ag^+ + NO_3^- + K^+ + Cl^- \longrightarrow AgCl\downarrow + K^+ + NO_3^-$$

Since mobility of potassium ions is nearly the same as that of silver ion, which it replaces, the conductance will remain more or less constant and will begin to increase only after the end point. Therefore on plotting a graph between conductance vs. volume of KCl added, the end point is obtained.

Procedure

1. Take 20 ml of the given $AgNO_3$ solution in a beaker. Dilute this solution by a known volume of conductivity water (Double distilled water) so that the electrodes of the cell are completely immersed in the solution.
2. Wash the cell with conductivity water and then rinse it with $AgNO_3$ solution. Connect the cell to the conductivity bridge.
3. Fix the cell in a holder.
4. Note the conductance of the $AgNO_3$ solution.

5. Rinse and fill the burette with N/10 KCl solution, preferably micro burette may be employed.
6. Add KCl solution from the burette into the solution of $AgNO_3$ (0.5 ml at a time). Note the conductance after each addition of 0.5 ml of KCl.

Carry on the experiment till 10 ml of KCl have been added.

Tabulation

Volume of $AgNO_3$ taken = 20 ml

No. of observations	*Volume of KCl added (ml)*	*Observed conductance (ohm^{-1})*
1	0.0 ml	
2	0.5 ml	
3	1.0 ml	
4	1.5 ml	
5	2.0 ml	
↓	↓	
21	10.0 ml	

Calculation

The values of observed conductance is plotted vs. volume of KCl added (Fig. 11.1).

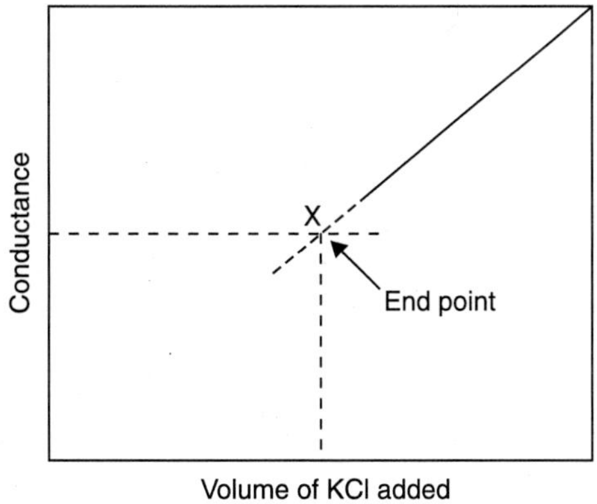

FIGURE 11.1 **Plot of conductance versus volume of KCl.**

The point of intersection gives the volume of KCl required for the complete precipitation of $AgNO_3$. By drawing a perpendicular from this point, volume V_2 will be obtained, which is required for complete precipitation of $AgNO_3$.

$$N_1V_1 = N_2V_2$$

$$\text{AgNO}_3 \qquad \text{KCl}$$

$$N_1 \times 20 = \frac{N}{10} \times V_2$$

$$N_1 \text{ (Normality of AgNO}_3) = \frac{N}{200} \times V_2$$

$$\therefore \quad \text{Stength} = \text{Eq. wt.} \times \text{Normality} = 143.5 \times N_1 = x \text{ g/L.}$$

Conclusion

The strength of given $AgNO_3$ solution = x g/L.

Preparation of Reagents

1. *Potassium Chloride Solution (N/10):* Dissolve 7.456 g of anhydrous KCl dried at 180°C for one hour in distilled water and make up the volume to 1 L.

11.4 POTENTIOMETRIC MEASUREMENT

Potentiometry deals with the measurement of difference in potential between two electrodes which have been combined to form electrochemical cells.

Potentiometer

In potentiometer the principle generally used is that of Poggendorff compensation method. The principle involved is shown in the Fig. 11.2. AB is the potentiometer slidewire of uniform cross-section and having high resistance. A storage cell 'C' is connected across the terminals of slidewire AB, such that potential drops from A to B.

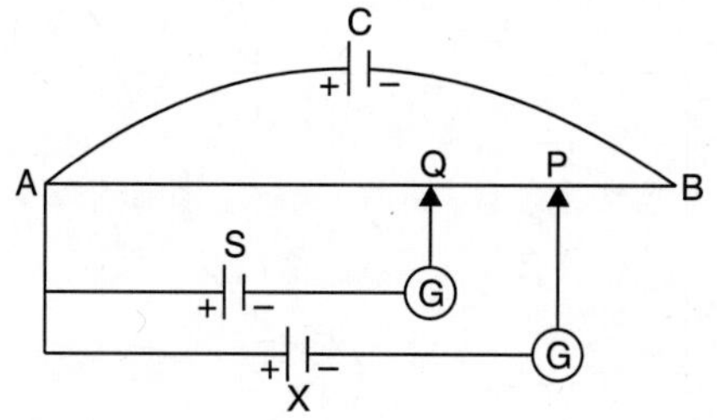

FIGURE 11.2 **Measurement of EMF of a cell.**

Now, the cell 'X' whose emf is required is connected to 'A'. The other terminal of cell 'X' is connected through a galvanometer (G) to a sliding contact P. This is moved along the slidewire until there is no deflection in the galvanometer. This means that the emf of cell 'X' just balances the drop of potential between A and P. Let the contact point is assumed to be at 'P' when there is no deflection. Next a standard cell 'S' of known emf is connected to A and thus replacing cell 'X'. The experiment is repeated. The sliding contact is again moved along the wire AB, till the point Q is reached, when there is no deflection.

If, E_s = emf of standard cell

E_x = emf of unknown cell

$E_s \propto$ Length AQ

$E_x \propto$ Length AP

$$\frac{E_x}{E_s} = \frac{\text{Length AP}}{\text{Length AQ}} \quad \text{or,} \quad E_x = \frac{\text{Length AP}}{\text{Length AQ}} \times E_s$$

The two lengths being known and since E_s the emf of standard cell is known, E_x is easily determined. The most widely used standard cell is the weston cell.

Potentiometric Titration

The determination of equivalence point of titrations on the basis of potential measurements is called potentiometric titrations. A good potentiometric titration requires a standard electrode and suitably selected indicator electrode. Determination of difference of potential of the indicator electrode can be used as indicator in volumetric titrations. The equivalence point is indicated by fairly a large change of potential. The potential of an indicator electrode goes on changing w.r.t. a standard electrode (calomel electrode) by the change of concentration of ions during titration.

Advantages of Potentiometric Titrations

1. These titrations are applicable to even coloured solutions.
2. These titrations give highly accurate results.
3. Weak acid-weak base titrations are possible by this method.
4. These are very rapid as compared with gravimetric methods.

EXPERIMENT 11.4

Aim of the Experiment

To estimate the strength of given hydrochloric acid solution by titrating against sodium hydroxide (0.1 N) using quinhydrone as the indicator electrode.

Apparatus Required

- Potentiometer
- Battery
- Calomel electrode
- Platinum electrode
- Beaker
- Stirrer (magnetic).

Chemicals required

- N/10 NaOH solution
- Quinhydrone
- Given hydrochloric acid solution.

Theory

In a potentiometric titration, a suitable electrode is immersed in the solution to be titrated acts as the indicator. The indicator electrode is paired with a reference electrode and the two electrodes are connected to an electronic voltmeter. Since the reference electrode potential has a constant value, any change in the indicator electrode potential is reflected by a similar change in the cell potential. Therefore the equivalence point can be found by plotting a graph between the cell emf and the volume of titrant added from the burette.

Procedure

Pipette out 20 ml of the given HCl solution in a 100 ml capacity beaker. Add a pinch of quinhydrone to the saturated solution. Connect the two electrodes, leading to a potentiometer and then measure the emf. Add 0.5 ml of alkali and measure the emf of the solution. Continue the addition, take 3–4 readings after the equivalence point. Record the volume of alkali added and emf of the solution.

Tabulation

Volume of acid solution taken = 20 ml

Normality of alkali used = 0.1 N

No. of observations	*Volume of alkali added (ml)*	*EMF of the cell (volts)*
1		
2		
3		

The strength of the acid can be calculated using the relation,

$$N_1V_1 = N_2V_2$$

N_1 = Normality of NaOH = 0.1 N

V_1 = Volume of the NaOH = x ml (say)

N_2 = Normality of the acid = ?

V_2 = Volume of the acid taken = 20 ml

$$N_2 = \frac{N_1V_1}{V_2} = \frac{0.1 \times x}{20} \text{ N}$$

Strength of HCl = Normality × equivalent weight

= N_2 × 36.5 g/L

Plot a graph by taking potentials along Y-axis and volume of NaOH added along x-axis. The volume corresponding to the steepest curve is the equivalence point.

Conclusion

Strength of the given acid solution is $\frac{0.1 \times x}{20} \times 36.5$ g/L

Preparation of Reagents

(*i*) *Sodium Hydroxide Solution (N/10):* Dissolve 4 g of NaOH in 1L of distilled water.

11.5 SPECTROPHOTOMETRIC MEASUREMENT

Spectroscopy may be defined as the interaction between the matter and electromagnetic radiations. It is the most powerful tool available in the analysis of wide range of samples.

Spectroscopic methods of analysis are based on measurements of the electromagnetic radiation emitted or absorbed by the solutes. When an electromagnetic radiation is passed through a sample (solid, liquid or gas), certain characteristic wavelengths are absorbed by the sample. The extent of absorption depends upon the wavelength of the radiation, thickness to the medium and also on the concentration of the sample. Thus, there is a decrease in the intensity of transmitted radiation. By measuring the decrease in intensity of radiation we can qualitatively and quantitatively analyse a given substance. The technique is known as photometry.

EXPERIMENT 11.5

Aim of the Experiment

(*a*) To determine λ_{max} (wavelength of maximum absorption) of solution of $KMnO_4$ using a spectrophotometer.

(*b*) Verify Beer's law and apply it to find the concentration of the given unknown solution.

Apparatus Required

- UV-visible spectrophotometer
- Beaker
- Tissue paper.

Chemicals Required

- Standard solution of $KMnO_4$
- Distilled water.

Theory

When an electromagnetic radiation is passed through a sample, certain characteristic wavelengths are absorbed by the sample. As a result the intensity of the transmitted light is decreased. The measurement of the decrease in intensity of radiation is the basis of spectrophotometry.

A spectrophotometer is a device which detects the percentage transmittance of light radiation when light of certain intensity and frequency range is passed through the sample. Thus the instrument compares the intensity of the transmitted light with that of incident light. The modern ultra violet-visible spectrometer consists of light source, monochromator, detector, amplifier and the recording devices (Fig. 11.3).

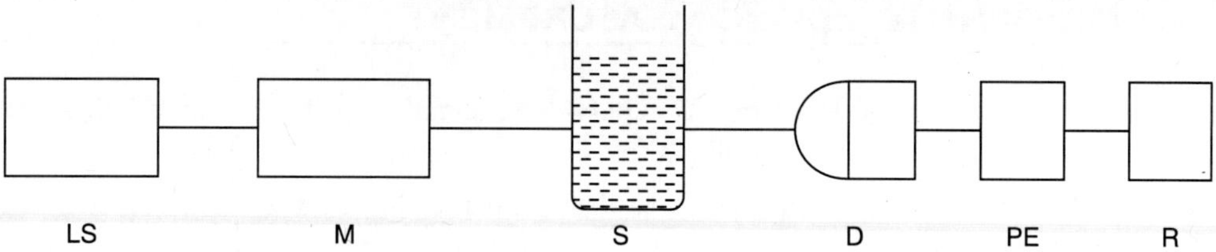

FIGURE 11.3 **UV-VIS spectrophotometer.**

When monochromatic light falls upon a homogeneous medium, a portion of the incident light is reflected, a portion is absorbed within the medium and the remainder is transmitted.

$$I_o = I_a + I_t + I_r$$

In case of aqueous solution, I_r is negligible as compared to I_a and I_t.

$$I_o = I_a + I_t$$

According to Beer Lamberts law the intensity of the incident light is proportional to the length of thickness of the absorbing medium and the concentration of the solution.

$$I_t = I_o \cdot 10^{-\varepsilon CL} \qquad \text{...(1)}$$

Where C is concentration of solute expressed in mole/L, L is the length of the cell and 'ε' is a constant characteristic of the solute called molar extinction coefficient or molar absorptivity. Further log I_o/I_t also called optical density (OD) or absorbance (A). Since absorbance 'A' of the medium is given by,

$$A = \log I_o/I_t \qquad \text{...(2)}$$

From eqn. (1) and (2)

$$A = \varepsilon CL$$

Transmittance 'T' of a solution is the ratio of I_t/I_o *i.e.*, the fraction of incident light transmitted by solution.

$$T = \frac{I_t}{I_o}$$

$$-\log T = \log \frac{I_o}{I_t} = A$$

$$A = -\log T = \varepsilon CL \qquad \text{...(3)}$$

Procedure

Setting of the Instrument

1. Ensure that meter initially reads zero on transmittance scale (T). If zero is not adjusted, set mechanical zero with adjusting knob.

2. Connect the instrument to the power supply and put on the power switch.
3. Adjust the wavelength knob to the required wavelength region on scale (approximately).
4. Choose the position of wave length switch correspondingly either to 340–400 nm or 400–960 nm.
5. Adjust the "Set Zero" knob so that meter needle reads Zero on T-scale and 100 on OD scale.

Working

6. Open the lid of the cell compartment and insert a cuvette containing blank solution *i.e.* distilled water. Close the lid properly.
7. Adjust the control knob (set 100) to 100% transmittance or zero optical density (OD).
8. Remove the cuvette and close the lid tightly again. Empty the cuvette and rinse it with the standard solution of $KMnO_4$. Fill it with standard solution.
9. Now place the cuvette containing the standard solution in the cell compartment. Note the OD and Transmittance.

Determination of λ_{max}

10. Now change the wavelength by 20 nm and note absorbance (OD) and transmittance for each wavelength.
11. Plot a graph between wavelength on the X-axis and absorbance (OD) on the Y-axis.

Verification of Beer's Law

12. Fix the wavelength at λ_{max} position.
13. Prepare $KMnO_4$ solutions with concentrations 0.2%, 0.5%, 1.0%, 1.5%, 2.0%, 2.5% and 3.0% etc. (20 ml each).
14. Note down the absrobance (OD) of series of solution of $KMnO_4$ prepared above by the method described above.
15. Plot a graph between OD against concentration (A straight line shows verification of Beer's law).
16. Now find out the OD of the unknown solution of the $KMnO_4$. Find out the concentration of this solution from the graph.

Tabulation

Determination of λ_{max}

No. of observations	*Wavelength (nm)*	*Absorbance (OD)*

Verification of Beer's Law

No. of observations	*Concentration (C) moles/L*	*Absorbance (OD)*

Calculation

1. A curve is plotted between wavelength and absorbance (OD). From the graph (Fig. 11.4) the maximum value of A gives λ_{max}.

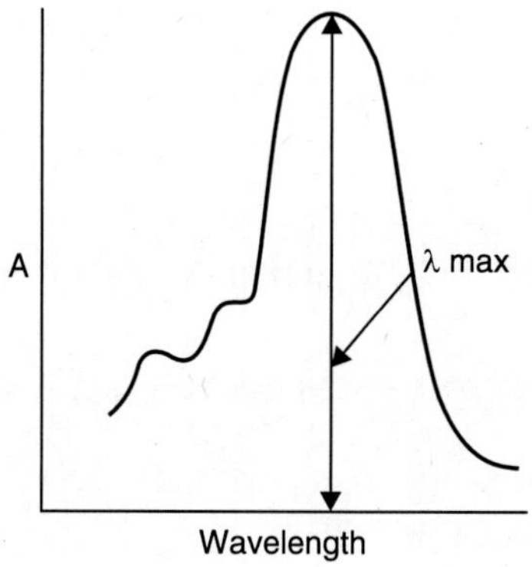

FIGURE 11.4

2. A curve is plotted between OD and concentration and if a straight line is obtained as shown by eqn. (3), Beer's law is verified (Fig. 11.5).

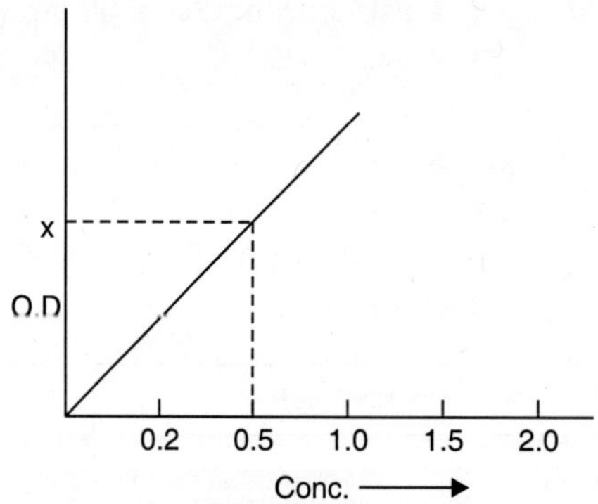

FIGURE 11.5

3. From the graph of OD versus concentration, the concentration of the unknown solution can be found out. For example, in the Fig. 11.5 if X is the OD of unknown solution then its concentration will be 1.0%.

Conclusion

(*a*) λ_{max} for $KMnO_4$ = ... nm

(*b*) $KMnO_4$ solution obeys Beer's law

(*c*) Concentration of the unknown solution = ... mg/L

EXPERIMENT 11.6

Aim of the Experiment

To determine ferrous ions in a given sample spectrophotometrically by O-phenanthroline method.

Apparatus Required

- Spectrophotometer
- Beaker
- Volumetric flask (50 ml).

Chemicals Required

- Hydroxylamine hydrochloride (10%)
- Phenanthroline solution (0.25%)
- Ammonium acetate buffer solution
- Sodium acetate solution (0.2%)
- Stock iron solution
- Standard iron solution
- Concentrated HCl.

Theory

Iron present in water in the ferric state is reduced to ferrous state by boiling with HCl and hydroxylamine hydrochloride. Then 1, 10-phenanthroline is added to the solution at pH 3.2–3.3 which forms a soluble chelated complex having orange-red colour with the ferrous ions present. The intensity of colour is independent of pH in the range 3 to 9. This can be used for the colorimetric or spectrophotometric determination of Fe^{2+} in the water sample.

Procedure

From 1% stock soluton of ferrous ion, prepare five standard solutions ranging from 0.1 to 0.5 normal concentrations. Transfer 1 ml of 0.1 N solution in a 50 ml volumetric flask. Add 5 ml each of 10% hydroxyl amine solution, sodium acetate solution and *o*-phenanthroline solution in order. Dilute the solution to 50 ml with distilled water and mix properly. Repeat the above process with another standards and unknown solution. Keep these coloured solutions at rest for five minutes. Note down the absorbance of these six solutions with spectrophotometer selecting 510–515 wavelength. Prepare a standard calibration curve by plotting absorbance against concentration of five standard solutions. The concentration of unknown solution, which corresponds to the absorbance, recorded earlier can be determined from the standard calibration curve.

Tabulation

No. of observations	*Concentration moles/L*	*Absorbance (OD)*

Conclusion

The concentration of ferrous ion in a given sample solution using spectrophotometer and by o-phenanthroline method is *x* mg/L.

Preparation of Reagents

(1) *Hydroxylamine Hydrochloride (10%):* Dissovle 10 g of NH_2OH . HCl in 100 ml distilled water.

(2) *Phenanthroline Solution (0.25%):* Dissolve 250 mg of 1, 10-phenanthroline monohydrate, $C_{12}H_8N_2 . H_2O$, in 100 ml of distilled water. Warm slightly or add 2 drops of conc. HCl 1 ml of this solution can chelate 100 mg of iron. Three molecules of phenanthroline chelate each atom of ferrous ion to form an orange red complex.

(3) *Ammonium Acetate Buffer Solution:* Dissolve 250 g of ammonium acetate in 150 ml of distilled water. Add 700 ml of glacial acetic acid.

(4) *Sodium Acetate Solution (0.2%):* Dissolve 200 mg of sodium acetate in 100 ml of distilled water.

(5) *Stock Iron Solution:* Dissolve 1.404 g of $Fe(NH_4)_2 . (SO_4)_2 . 6\,H_2O$ in 50 ml distilled water to which 20 ml conc. H_2SO_4 is added earlier. Add 0.1 N $KMnO_4$ solution dropwise till a faint pink colour persists. Dilute to 1 litre.

(6) *Standard Iron Solution:* Dilute 50 ml of the above stock iron solution to 1 litre. This solution should be prepared freshly. 1 ml of this solution corresponds to 10.00 μg of Fe.

VIVA-VOCE

1. Define the term pH.
2. What are buffer solutions?
3. How the equivalence point in acid base titrations is determined pH metrically?
4. What is the effect of temperature on pH?
5. What are calomel and glass electrode?
6. Define conductance.

   Ans. Conductance is the ease of flow of current through a solution of electrolyte. This is the reciprocal of resistance (R) and has the unit ohm^{-1} or mho.

7. What is conductivity water?
8. What is the effect of temperature on conductance?

   Ans. The conductance of solution increases with increase in temperature. This is due to the

   (*i*) increase in velocity of ions

   (*ii*) decrease in viscosity of the medium

   (*iii*) decrease in interaction between the ions.
9. What is conductivity of water?
10. What is e.m.f. of a cell?
11. What is electrochemical cell?

    Ans. A device used to convert the chemical energy produced in a redox reaction into electrical energy is called an electrochemical cell or galvanic cell or voltaic cell.
12. In an electrochemical cell which electrode is cathode and which electrode anode.
13. What is a reference electrode? Give some examples.
14. What is spectrophotometry?
15. What is absorbance?
16. What is the structure of *o*-phenanthroline complex with iron (II)?

    Ans.

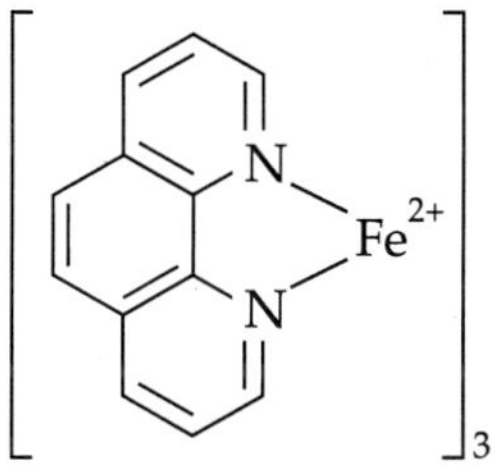

$(Phen)_3\ Fe^{2+}$

This is called as Ferroin.

SOLVED NUMERICAL PROBLEMS

1. Calculate the molarity of NaOH solution, 2 g of which have been dissolved in 500 cm^3 solution.

   Solution.

$$\text{Mass of NaOH} = 2\ g$$

$$\text{Molecular mass of NaOH} = 40$$

$$\text{Volume of solution} = 500\ cm^3$$

$$500\ cm^3 \text{ solute contain NaOH} = 2\ g$$

$$1000\ cm^3 \text{ solute contain NaOH} = \frac{2}{500} \times 1000 = 4\ g$$

$$\therefore \quad \text{Molarity of solution} = \frac{4}{40} = 0.1\ M$$

2. Find the molarity and molality of a 15% solution of H_2SO_4 (density of H_2SO_4 = 1.10 g/ml and mol. mass = 98).

Solution.

(*i*) *To find molality of solution*

Amount of H_2SO_4 present in 100 g of H_2SO_4 solution = 15 g

Amount of water = 85 g

Amount of H_2SO_4 present in 1000 g of $H_2O = \dfrac{15}{85} \times 1000$ g = 176.47 g

No. of moles of H_2SO_4 present in 1000 g of water $= \dfrac{176.47}{98} = 1.8$

∴ Molality = 1.8 m

(*ii*) *To find molarity of solution*

15 g H_2SO_4 is present in 100 g H_2SO_4 solution

Volume of 100 g solution $= \dfrac{100}{1.1}$ ml

$\dfrac{100}{1.1}$ ml of solution contains H_2SO_4 = 15 g

1000 ml of solution contains $H_2SO_4 = \dfrac{15 \times 1.1}{100} \times 1000 = 165$ g

No. of moles of sulphuric acid present in 1000 ml of sulphuric acid solution $= \dfrac{165}{98} = 1.68$

∴ Molarity = 1.68 M.

3. Find the molality of H_2SO_4 solution whose specific gravity is 1.98 and contains 95 g of H_2SO_4 per 100 ml.

Solution.

Weight of 1 L solution = Volume in ml × Specific gravity

= 1000 × 1.98

= 1980 g

Weight of H_2SO_4 in 1 L solution = 95 × 10 = 950 g

Weight of water in the solution = 1980 – 950 = 1030 g

$$\text{Molality, } m = \frac{W_B}{m_B \times W_A} \times 1000$$

where W_B = Weight of solute

m_B = Molecular weight of solute

W_A = Weight of solvent.

$$m = \frac{950}{98 \times 1030} \times 1000$$

$$= 9.415 \text{ m.}$$

4. The density of a 2.03 M solution of acetic acid in water is 1.017 g/mole. Calculate the molality of the solution.

Solution.

Weight of one litre solution $= 1000 \times 1.017$

$= 1017$ g

Weight of acetic acid in 1 L solution

$= \text{Molarity} \times \text{Molecular weight}$

$= 2.03 \times 60$

$= 121.8$ g

Weight of water $= 1017 - 121.8 = 895.2$ g

$$\text{Molality} = \frac{W_B}{m_B \times W_A} \times 1000$$

$$= \frac{121.8}{60 \times 895.2} \times 1000 = 2.2676 \text{ m.}$$

5. Calculate the molar concentration of pure water at 25°C. (Density of water is 1 g/ml).

Solution.

$$\text{Molar concentration of water} = \frac{\text{Weight of one litre water}}{\text{Molecular weight of water}}$$

$$= \frac{1000}{18} = 55.55 \text{ M.}$$

6. Calculate molar volume of the 0.02 M $Na_2S_2O_3$ solution of density 1.10 g/L.

Solution.

Moles of solute $= 0.02$

Moles of solute $= 0.02 \times 158 = 3.16$

Weight of 1L solution $= 1000 \times 1.10 = 1100$ g

W_A (weight of solvent) $= 1100 - 3.16$

$= 1096.84$

$$\text{Moles of solvent} = \frac{W_A}{m_A} = \frac{1096.84}{18} = 60.935$$

Total moles in the solution $= 60.935 + 0.02 = 60.955$ mole

$$\text{Molar volume} = \frac{1000}{60.955} = 16.40 \text{ ml/mole.}$$

7. The density of a 26% solution of KOH is 1.24 g/ml. How many moles of KOH are contained in 5 L of the solution?

Solution.

We know that molarity, $= \dfrac{x \times d \times 10}{\text{Molecular weight}}$

where x = % by weight

d = density

$$\therefore \quad \text{Molarity} = \frac{26 \times 1.24 \times 10}{56} = 5.757$$

Thus, 5 L of solution will contain 5 × 5.757 mole of KOH = 28.785 moles

8. Calculate the normality of $Ba(OH)_2$ solution if it takes 42 ml of 4M HCl to neutralise 31.5 ml of the base.

Solution.

We know that,

$$\underset{\text{Acid}}{N_1V_1} = \underset{\text{Base}}{N_2V_2}$$

$$42 \times 4 = N_2 \times 31.5$$

$$N_2 = 5.333.$$

9. Find the normality of a solution obtained by mixing 100 ml of 0.2 N HCl and 500 ml of 0.12 M H_2SO_4.

Solution.

No. of equivalents in 100 ml of 0.2 N HCl

$$= \frac{0.2}{1000} \times 100 = 0.02$$

No. of equivalents in 500 ml of 0.12 M H_2SO_4

$$= \frac{0.24}{1000} \times 500 = 0.12$$

Here, normality of H_2SO_4 $= 0.12 \times 2 = 0.24$

Total equivalents $= 0.02 + 0.12 = 0.14$

Total volume after mixing $= 600$ ml

$$\text{Normality} = \text{No. of equivalents per litre}$$

$$= \frac{0.14}{600} \times 1000$$

$$= 0.233 \text{ N.}$$

10. Calculate the molality of 1 litre solution of 93% H_2SO_4 (weight/weight). The density of the solution is 1.84 g/ml.

Solution.

Weight of one litre solution $= 1000 \times 1.84$

$= 1840$ gm.

Weight of H_2SO_4 in solution of one litre

$$= \frac{93}{100} \times 1840$$

$= 1711.2$ gm.

Weight of water $= 1840 - 1711.2 = 128.8$

$$\text{Molality } (m) = \frac{W_B}{m_B \times W_A} \times 1000$$

where $B \rightarrow$ solute, $A \rightarrow$ solvent

$$m = \frac{1711.2}{98 \times 128.8} \times 1000$$

$= 135.56.$

11. A sample of spirit has 90% ethanol by weight. Calculate the mole fraction of ethanol and water.

Solution.

Weight of ethanol $= 90$ gm.

Weight of water $= 10$ gm.

$$\text{Moles of ethanol} = \frac{\text{Weight}}{\text{Molecular weight}} = \frac{90}{46}$$

$$\text{Moles of water} = \frac{10}{18}$$

$$\text{Mole fraction of ethanol} = \frac{nC_2H_5OH}{nC_2H_5OH + nH_2O} = \frac{1.956}{1.956 + 0.555} = 0.779$$

Mole fraction of water $= 1 - 0.779 = 0.221.$

12. 2.82 g of glucose (mol. mass = 180) are dissolved in 30 g water. Calculate the molality of the solution?

Solution.

Weight of glucose $= 2.82$ g

Molecular mass of glucose $= 180$

$$\therefore \text{ Number of moles of glucose} = \frac{2.82}{180}$$

Weight of solvent water $= 30$ g

We know that,

$$\text{Molality of glucose solution} = \frac{\text{No. of moles of glucose}}{\text{Wt. of solvent}} \times 1000$$

$$= \frac{2.82}{180} \times \frac{1000}{30} = 0.52 \text{ molal.}$$

13. 200 ml of a solution contains 0.754 g sodium bicarbonate $NaHCO_3$. What is its normality?

Solution. Amount of $NaHCO_3$ in 200 ml solution = 0.754 g

∴ Amount of $NaHCO_3$ in 1000 ml solution = 0.754 × 5 = 3.77 g

Molecular mass of $NaHCO_3$ = 23 × 1 + 12 + 3 × 16

= 83

Equivalent mass of $NaHCO_3$ $= \frac{\text{Molecular mass}}{\text{Acidity}} = \frac{83}{1} = 83$

∴ $$\text{Normality} = \frac{\text{Amount in grams per litre}}{\text{Equivalent mass}} = \frac{3.77}{83}$$

= 0.0454217.

14. How many ml of 1M H_2SO_4 would be required to neutralise 100 ml of 1N NaOH solution?

Solution.

We know that

x cc. of 1M H_2SO_4 = x cc. of 2N H_2SHO_4

x cc. of 2N H_2SO_4 = 100 cc. of 1N NaOH

Applying normality equation,

$$\underset{\text{Acid}}{N_1V_1} = \underset{\text{Alkali}}{N_2V_2}$$

$$2N \times x = 1\,N \times 100$$

∴ $$x = 50$$

Thus, volume of 1M H_2SO_4 required = 50 cc or ml.

15. 40 ml of sodium hydroxide solution of unknown concentration was neutralised by the addition of 30 ml of 0.2 N sulphuric acid. What is the normality of sodium hydroxide?

Solution.

Let the normality of NaOH solution be N

Applying normality equation

$$\underset{(NaOH)}{N_1V_1} = \underset{(H_2SO_4)}{N_2V_2}$$

$$N \times 40 = 0.2 \times 30$$

$$N = \frac{30 \times 0.2}{40} = 0.15\,N$$

Normality of sodium hydroxide is 0.15 N.

16. Find out the weight of NaOH required to prepare 2 L of N/10 solution.

Solution.

Equivalent weight of NaOH = 40

One litre of 1N solution contain NaOH = 40 g.

Two litre of $\frac{N}{10}$ solution contain NaOH = $40 \times \frac{1}{10} \times 2 = 8$ g.

17. Calculate the weight of NaOH required to neutralise 25 ml of 1N H_2SO_4.

Solution.

1000 ml of 1 N H_2SO_4 = 1 g equivalent of H_2SO_4

$$25 \text{ ml of 1N } H_2SO_4 = \frac{1}{1000} \times 25 = \frac{1}{40} \text{ g equivalent of } H_2SO_4$$

$$\frac{1}{40} \text{ g. eq. of } H_2SO_4 = \frac{1}{40} \text{ g eq. of NaOH}$$

$$= \frac{1}{40} \times 40 = 1 \text{ gm of NaOH.}$$

18. Density of air is 0.001293 g/ml. What is its molecular weight?

Solution.

1 gm mole of a gas occupies 22400 ml

Thus, molecular weight of air = density × 22400

= 0.001293 × 22400

= 28.96 gm/mole.

19. 2 g of NaOH are present in 800 ml of solution. Find its normality.

Solution.

Equivalent weight of NaOH = 40

One litre of 1N solution contain NaOH = 40 g.

To find out the normality of 2g of NaOH in 800 ml solution,

$$W = \frac{ENV}{1000}$$

W = weight of NaOH

E = Eq. weight

N = Normality

V = Volume of the solution

$$2 = \frac{40 \times N \times 800}{1000}$$

N = 0.0625.

20. Calculate the number of sulphate ions in 100 ml of 0.001 M H_2SO_4.

Solution.

Weight of H_2SO_4 $= \frac{\text{Molarity}}{1000} \times \text{Volume} \times \text{Molecular weight}$

$$= \frac{0.001}{1000} \times 100 \times 98 = 9.8 \times 10^{-3} \text{ g}$$

$$H_2SO_4 \leftrightarrow 2H^+ + SO_4^{2-}$$

$\because$ 98 g H_2SO_4 contains 96 g SO_4^{2-} ions

$\therefore$ 9.8×10^{-3} g H_2SO_4 contains $= \frac{96}{98} \times 9.8 \times 10^{-3} = 9.6 \times 10^{-3}$ g.

21. Density of water at room temperature is 1 gm/ml. How many molecules are there in one drop of water if its volume is 0.05 cm^3?

Solution.

Weight of one drop of water = $0.05 \times 1 = 0.05$ gm

No. of molecules of water in one drop

$$= \frac{\text{Weight}}{\text{Molecular weight}} \times N$$

$$= \frac{0.05}{18} \times 6.023 \times 10^{23}$$

$$= 0.0167 \times 10^{23} = 1.67 \times 10^{21} \text{ molecules.}$$

22. Calculate the number of $C_2O_4^{2-}$ ions in 100 ml. of 0.2 M aqueous $Na_2C_2O_4$. Assume complete ionisation?

Solution.

No. of moles of $Na_2C_2O_4$ in 100 ml of 0.2 M solution

$$= 100 \times 10^{-3} \times 0.2 = 0.02$$

No. of moles of $C_2O_4^{2-}$ ions furnished by 0.02 moles of $Na_2C_2O_4$ = 0.02 moles.

No. of oxalate ions $= 0.02 \times 6.02 \times 10^{23} = 1.24 \times 10^{22}$.

23. If H_2SO_4 ionises as

$$H_2SO_4 + 2H_2O \longrightarrow 2H_3O^+ + SO_4^{2-}$$

Then what is the total number of ions produced by 0.1 M H_2SO_4?

Solution.

1 M H_2SO_4 produces ions = 3 moles

0.1 M H_2SO_4 produces ions = 0.3 mole

No. of ions $= 0.3 \times 6.02 \times 10^{23} = 1.8 \times 10^{23}$.

24. 'M' is the molecular mass of $KMnO_4$. What is the equivalent mass of $KMnO_4$ when it is converted itno K_2MnO_4?

Solution.

$$2KMnO_4 + 2KOH \longrightarrow 2\,K_2MnO_4 + H_2O + [O]$$

$$\overset{+7}{Mn}O_4^- + e^- \longrightarrow \overset{+6}{Mn}O_4^{2-}$$

$\therefore$ Equivalent mass of $KMnO_4$ when it is converted into K_2MnO_4

$$= \frac{\text{Mol. mass}}{\text{Change in oxidation number}} = \frac{M}{1} = M.$$

APPENDICES

APPENDIX I

Periodic Table of Elements

H 1																	He 2
Li 3	Be 4											B 5	C 6	N 7	O 8	F 9	Ne 10
Na 11	Mg 12											Al 13	Si 14	P 15	S 16	Cl 17	Ar 18
K 19	Ca 20	Sc 21	Ti 22	V 23	Cr 24	Mn 25	Fe 26	Co 27	Ni 28	Cu 29	Zn 30	Ga 31	Ge 32	As 33	Se 34	Br 35	Kr 36
Rb 37	Sr 38	Y 39	Zr 40	Nb 41	Mo 42	Tc 43	Ru 44	Rh 45	Pd 46	Ag 47	Cd 48	In 49	Sn 50	Sb 51	Te 52	I 53	Xe 54
Cs 55	Ba 56	*	Hf 72	Ta 73	W 74	Re 75	Os 76	Ir 77	Pt 78	Au 79	Hg 80	Ti 81	Pb 82	Bi 83	Po 84	At 85	Rn 86
Fr 87	Ra 88	#	Ku 104	Ha 105													

* Lanthanides			La 57	Ce 58	Pr 59	Nd 60	Pm 61	Sm 62	Eu 63	Gd 64	Tb 65	Dy 66	Ho 67	Er 68	Tm 69	Yb 70	Lu 71
# Actinides			Ac 89	Th 90	Pa 91	U 92	Np 93	Pu 94	Am 95	Cm 96	Bk 97	Cf 98	Es 99	Fm 100	Md 101	No 102	Lr 103

APPENDIX II

List of Equivalent Weights of Substances Commonly Used in Volumetric Analysis:

Titration and substances used	*Formula*	*Eq. weight*
Acid alkali titration:		
Sodium carbonate	Na_2CO_3	M.W./2 = 53.06
Sodium bicarbonate	$NaHCO_3$	M.W./1 = 84.00
Potassium carbonate	K_2CO_3	M.W/2 = 69.00
Potassium bicarbonate	$KHCO_3$	M.W./1 = 100.00
Sodium hydroxide	$NaOH$	M.W./1 = 40.00
Barium hydroxide	$Ba(OH)_2.8H_2O$	M.W./2 = 157.75
Oxalic acid (hydrated)	$H_2C_2O_4.2H_2O$	M.W./2 = 63.04
Hydrochloric acid	HCl	M.W./1 = 35.46
Sulphuric acid	H_2SO_4	M.W./2 = 49.04
Redox titrations		
Pot. permanganate	$KMnO_4$	M.W./5 = 31.606
Pot. dichromate	$K_2Cr_2O_7$	M.W./6 = 49.03
Ferrous sulphate (hydrated)	$FeSO_4.7H_2O$	M.W./1 = 278.00
Ferrous sulphate (anhydrous)	$FeSO_4$	M.W./1 = 152.00
Ferrous ammonium sulphate	$FeSO_4.(NH_4)_2SO_4$ $6H_2O$	M.W./1 = 392.14
Oxalic acid (hydrated)	$H_2C_2O_2.2H_2O$	M.W./2 = 63.04
Iron (ferrous)	Fe	At. W./1 = 55.84
Iodine titrations		
Iodine	I_2	At.W./1 = 126.92
Hypo (Sod. thiosulphate)	$Na_2S_2O_3.5H_2O$	M.W./1 = 248.19
Hydrated		
Anhydrous	$Na_2S_2O_3$	M.W./1 = 158.09
Arsenious oxide	As_2O_3	M.W./4 = 49.45
Copper sulphate (hydrated)	$CuSO_4.5H_2O$	M.W./1 = 249.68
Argentometric titrations		
Silver nitrate	$AgNO_3$	M.W./1 = 169.89
Silver	Ag	At.W./1 = 107.88
Sodium chloride	$NaCl$	M.W./1 = 58.45
Chloride ion	Cl^-	At.W./1 = 35.46
Pot. sulphocyanide	$KCNS$	M.W./1 = 97.17
Amm. sulphocyanide	NH_4CNS	M.W./1 = 76.12

APPENDIX III

Data on the Specific Gravity, Percentage Composition, Normality etc. of Some Reagents

(a) Concentrated Reagents

Reagents	*Specific gravity*	*Percentage by weight*	*Approximate normality*	*Weight of anhydrous reagent (gm) in 1 c.c.*
H_2SO_4	1.84	96	36N	1.80
HCl	1.19	37.9	12N	0.366
HNO_3	1.42	69.8	16N	0.991
NH_4OH	0.882	34.95	15N	0.308
Glacial CH_3COOH	1.055	99.5	17N	1.80

(b) Dilute Reagents

Reagents	*Specific gravity*	*Normality (approximate)*	*Anhydrous reagent (gm) in 1 ml*	*Preparation*
				Add
CH_3COOH	1.0256	3N	—	180 ml conc. acid to 820 ml H_2O
H_2SO_4	1.15	5N	0.240	1 ml conc. acid to 7 ml H_2O
HCl	1.05	3N to 5N	0.104	3 ml conc. acid to 7 ml H_2O
HNO_3	1.165	5N	—	3 ml conc. acid to 10 ml H_2O
NH_4OH	0.96	5N	0.089	4 ml conc. soln. to 10 ml H_2O
NaOH	1.113	3N	0.110	10 gm NaOH to 90 ml water
$Na_2CO_310H_2O$	1.079	1.5N	0.080	20 gm decahydrate to 80 ml water
$BaCl_2$	1.095	N	0.109	10 gm $BaCl_2.2H_2O$ to 100 ml water

APPENDIX IV

List of Acid Base Indicators

S. No.	*Indication*	*Acid colour*	*Base colour*	*pH range*
1.	Alizarin yellow	Yellow	Lilac	10.1–11.1
2.	Bromocresol green	Yellow	Blue	3.8–5.4
3.	Bromothymol blue	Yellow	Blue	6.8–8.4
4.	Malachite green (acidic)	Yellow	Blue green	0–1.8
5.	Malachite green (base)	Green	Colourless	11.4–13.0
6.	Methyl orange	Red	Yellow orange	3.1–4.6
7.	Methyl red	Red	Yellow	4.4–6.2
8.	Methyl violet	Yellow	Violet	0–2
9.	Thymol blue (acidic)	Red	Yellow	1.2–2.8
10.	Thymol blue (base)	Yellow	Blue	8.0–9.8
11.	Phenol red	Yellow	Red	6.8–8.4
12.	Phenolphthalein	Colourless	Red	8.2–9.8
13.	Thymolphthalein	Colourless	Blue	9.3–10.5
14.	Bromophenol blue	Yellow	Blue	3.0–4.6
15.	Congo red	Blue	Red	3.0–5.0
16.	Cresol red (acid)	Red	Yellow	0.2–1.8
17.	Cresol red (base)	Yellow	Red	7.2–8.8

APPENDIX V

Strength of Some Aqueous Solutions

Reagent	*% by mass*	*Specific gravity*	*Approximate normality*	*Approximate molarity*	*Volume required for 1 M per litre of solution*
Hydrochloric acid HCl	33%	1.16	12 N	12 M	84 ml
Sulphuric acid H_2SO_4	98%	1.84	36 N	18 M	56 ml
Nitric acid HNO_3	67%	1.42	16 N	16 M	63 ml
Phosphoric acid, H_3PO_4	85.5%	1.70	44.4 N	14.8 M	68 ml
Acetic acid	99.8%	1.05	17.45 N	17.45 M	57 ml
Perchloric acid	70.5%	1.67	11.7 N	11.7 M	86 ml
Formic acid	90.5%	1.20	23.6 N	23.6 M	43 ml
Ammonia (Ammonium hydroxide)	56.6%	0.90	14.5 N	14.5 M	69 ml

APPENDIX VI

Flash Point Ranges of Petroleum Fractions

Petroleum fractions	*Boiling point range (°F)*	*Flash point (by closed cup apparatus) (°F)*
Petroleum crude		20–90
Petroleum ether	100–160	– 50
Gasoline	100–400	– 40
Naphtha	300–400	100–110
Kerosene	350–550	110–120
Diesel	400–600	130–150
Gas oil	400–750	+ 150
Lubricating oil		
1. Spindle	—	170
2. Lingth Machine	—	320
3. Turbine	—	400
4. Air-craft engine	—	20–470
5. Motor	—	450
6. Cylinder	—	535
7. Transformer oil	550–750	340

APPENDIX VII

Drying Agents

Substance used	*Remarks*
$CaSO_4$	Very efficient drying agent and is converted to its hydrate, $2CaSO_4 . H_2O$
$MgSO_4$	Forms heptahydrate, $MgSO_4 . 7H_2O$
Na_2SO_4	Its efficiency is not so high.
$CaCl_2$	Fairly good capacity of dehydrating. It cannot be used to dry amines, alcohols and phenols.
Metallic sodium	Used in the form of wire for drying some organic solvents.

Index